Biological Communities Respond to Multiple Human-Induced Aquatic Environment Change

Biological Communities Respond to Multiple Human-Induced Aquatic Environment Change

Special Issue Editors

Marina Manca
Roberta Piscia

MDPI • Basel • Beijing • Wuhan • Barcelona • Belgrade • Manchester • Tokyo • Cluj • Tianjin

Special Issue Editors

Marina Manca   Roberta Piscia
CNR IRSA   CNR IRSA
Italy   Italy

Editorial Office
MDPI
St. Alban-Anlage 66
4052 Basel, Switzerland

This is a reprint of articles from the Special Issue published online in the open access journal *Water* (ISSN 2073-4441) (available at: https://www.mdpi.com/journal/water/special_issues/aquatic_environment_change).

For citation purposes, cite each article independently as indicated on the article page online and as indicated below:

LastName, A.A.; LastName, B.B.; LastName, C.C. Article Title. *Journal Name* **Year**, *Article Number, Page Range*.

ISBN 978-3-03928-544-0 (Pbk)
ISBN 978-3-03928-545-7 (PDF)

Cover image courtesy of Walter Zerla.

Contents

About the Special Issue Editors

Marina Manca is presently Senior Research Associate at CNR IRSA, Verbania, and was previously Senior Researcher at CNR ISE (Institute for Ecosystem Study, formerly "Istituto Italiano di Idrobiologia"). Between 2014 and 2016, she was director of CNR ISE. Her fields of interest include ecology and population dynamics of freshwater zooplankton; reconstruction of zooplankton communities from fossil remains of lake sediments; functional ecology and long-term changes in freshwater zooplankton communities under human and climate impacts; changes in the diversity of zooplankton communities and the role of lake egg banks; carbon and nitrogen stable isotope analyses for investigating lacustrine food webs and infrazooplankton predation.

Roberta Piscia has a degree in Biological Sciences and PhD in Environmental Sciences and is presently a permanent technician at CNR—Water Research Institute (Verbania), and has been working at CNR—Institute for Ecosystem Study since 2009. Her fields of expertise include ecology of freshwater zooplankton; analysis of zooplankton resting stages in lacustrine sediments; analyses of freshwater planktonic food webs and the role of crustacean pelagic zooplankton in the transfer of persistent organic pollutants by means of carbon and nitrogen stable isotope analyses.

Preface to "Biological Communities Respond to Multiple Human-Induced Aquatic Environment Change"

Perturbations linked to the direct and indirect impacts of human activities during the Anthropocene affect the structure and functioning of aquatic ecosystems to varying degrees. Some of these events stress aquatic life, including soil and water acidification, soil erosion, loss of base cations, release of trace metals or organic compounds, and application of essential nutrients capable of stimulating primary productivity. Superimposed on these changes, climate warming directly affects aquatic environments via altering species' metabolic processes, and indirectly by modifying food web interactions. Many stressors interact in a manner that can be difficult to predict. In part, this difficulty arises from the different possible responses by species or entire taxonomic groups to stressors, which may interact additively, synergistically, or antagonistically. Entire food webs may restructure if different trophic levels have consistently different responses to climate warming. However, the consequences of warming-induced changes in the food web structure for the long-term population dynamics of different trophic levels remain poorly understood. Such changes may be particularly important to understand in lakes, where food web production is socio-economically important and most organisms are ectotherms that are highly sensitive to changes in their surrounding environment. To understand the degree and mechanisms through which stressors affect lake biological communities and alter ecosystem functioning, long-term analyses by means of contemporary and paleo data are essential. Due to its remarkable physical inertia, including thermal stability, global warming is expected to also have a profound effect on groundwater ecosystems. The degree to which alterations affect aquatic ecosystem structure and functioning also requires addressing functional diversity at the molecular level to reconstruct the role different species play in the transfer of material and energy through the food web. In this Special Issue, we present examples of the impact of different stressors on aquatic ecosystems and their interactions, providing long-term, metabolic, molecular, and paleolimnological analyses.

In the last decades, anthropogenic activity, such as intensive animal farming and the use of fertilizers, has increased the inputs of nitrogen, affecting water quality and aquatic biodiversity, and promoting proliferation and toxicity of Cyanobacteria, with considerable socioeconomic consequences. In the paper by Leone et al., the efficiency and the transformation velocity of ammonium into oxidized compounds in south alpine lakes in Northern Italy of similar origin but differing in trophic status are compared in laboratory-scale experiments, performed in artificial microcosms. The important roles of total phosphorus and nitrogen concentrations on the commencement of the oxidation process is highlighted. The rate of nitrification was found to vary with natural concentration of ammonium in the studied lakes.

Increase in global climate variability (e.g., precipitation) is expected to result in more extreme hydrological events, such as extended droughts and flooding. Thus, studies on the ecological consequences of excessive water level fluctuations in lakes and reservoirs are receiving increased interest to predict the impact of climate on water ecosystem quality and health. Pan et al. assessed the changes in phytoplankton assemblages in response to water level increase in Danjiangkou Reservoir, one of the largest drinking water reservoirs in Asia. By applying non-metric multi-dimensional scaling analysis, the authors found that after the water level increases, seasonal variation in phytoplankton assemblages, with diatom dominance in both early and late seasons, was less evident, proving that water level increases alter the natural dynamics of the phytoplankton. This result suggests a non-negligible impact on the ecosystem of Danjiangkou Reservoir.

Functional-based approaches are being increasingly used to study aquatic ecosystems as an alternative to traditional taxonomy-based approaches. Functional diversity is a biodiversity measure based on the ecological role of the species present in a community. Visconti et al. studied the use of δ13C and δ15N stable isotopes as a proxy of zooplankton functional traits in Lake Maggiore, a large, deep subalpine Italian lake. Seasonal pattern of δ13C and δ15N signatures of different crustacean zooplankton taxa allowed the tracing of food sources, preferred habitats, and trophic positions of crustacean zooplankton taxa throughout one year. Selective vs. non-selective food habits as well as infra-zooplankton predator–prey interactions were clearly identified, highlighting the role of seasonality in shaping pelagic food webs and trophic interactions.

A comparative study of carbon and nitrogen stable isotope analyses of zooplankton of five subalpine lakes sampled in spring and summer along a trophic gradient (from oligotrophy to hyper-eutrophy) by Piscia et al. highlights some general patterns of carbon and nitrogen stable isotope zooplankton signatures. Taxa-specific isotopic signatures and changes with the season were different in shallow vs. deep lakes, and nitrogen isotopic signature reflected lake trophic status. Taxa-specific analyses within zooplankton community appear essential for understanding trophic relationships, changes in habitat, and carbon sources fueling the pelagic food web.

In combination with intensified climate warming, nutrient enrichment of freshwaters is a worldwide challenge, threatening aquatic biodiversity and ecosystem services. To understand impact of these perturbations on aquatic ecosystems and their functioning, paleolimnological studies are increasingly used for understanding relationships among functional diversity and ecosystem productivity, climate change, and trophic dynamics. The paper by Nevalainen et al. provides a detailed reconstruction of changes in the Cladocera community and its functional assemblages during eutrophication and its reversal in a large, deep subalpine lake (Lake Maggiore, Italy) using sedimentary records. By applying multivariate analysis techniques, the authors highlight the importance of bottom-up controls (i.e., total phosphorus) for shaping functional assemblages and top-down control by predators, particularly the predaceous cladoceran *Bythotrephes longimanus* for taxonomic community changes. Cladoceran functionality therefore is proved to be related to fundamental ecosystem functions, such as productivity, providing insights for long-term changes in ecological resilience.

Due to climate change, economic development, and population growth, approximately four billion people of the world's population, with nearly half of them living in India and China, are facing severe water scarcity, in terms of both quantity and quality. Assessing water quality and the ecological integrity of aquatic ecosystems is essential in this respect, requiring accurate and rapid measurements. Gao et al. compared traditional optical microscopy methods (TOM) and DNA barcoding (the 16S and 18S rRNA high-throughput sequencing method, HTS) for water bioassessment of the Danjiangkou Reservoir, the largest drinking water source in China, affecting more than 53 million people in Beijing and other receiving-water regions. The study highlights how differences between the two methods vary among stations. Overall, the study suggests a high reproducibility and potential for standardization and parallelization, supporting DNA barcoding as an excellent candidate for the simultaneous monitoring of plankton assemblages, including both phytoplankton and bacterioplankton for accurate and rapid monitoring of drinking water quality.

Climate change and enhanced nutrient loading are likely to stimulate the development of harmful Cyanobacteria blooms (cyanoHABs) affecting use, safety, and sustainability of water resources, resulting in considerable ecological and socioeconomic costs. Nava et al. report a study on *Planktothrix rubescens* and *Tychonema bourrellyi*, two potentially harmful Cyanobacteria in the deep South-Alpine Lake Iseo. A temporal shift in their development of the two species, linked to different capacities for overcoming winter and mixing periods, highlights the important role of the stability of the water column in determining *T. bourrellyi* settlement in Lake Iseo and the role of solar radiation in spring population development. The study confirms modified lake hydrodynamics due to climate change as the key factor for understanding occurrence of cyanoHABs and the increasing success of allochthonous Cyanobacteria in lakes all over the world.

Groundwater will play a fundamental role in sustaining ecosystems and enabling human adaptation to climate change. The strategic importance of groundwater will intensify as climate extremes become more frequent and intense. The effects of global warming on groundwater chemistry, hydro-geophysical properties and resources are relatively well known; studies of the biological responses to groundwater temperature increase are lacking. Di Lorenzo and Galassi provide an example of how temperature increase consistent with the foreseen increase in the next 30 years due to global warming will impact the physiology of *Diacyclops belgicus* (Kiefer, 1936), an obligate groundwater species with a wide geographic distribution. Controversial results of the experimental study do not provide full certainty about the response to global warming. However, the probable beginning of an irreversible denaturation of enzymes/proteins at a temperature increase of 10 °C from the thermal optimum poses serious warning for the fate of this species under global warming scenario.

Largely distributed in various freshwater habitats, freshwater oligochaetes are widely applied as indicator species in environmental assessment. Most studies, however, focus on their taxonomy, whereas relationships between the distribution of oligochaetes and their habitats are still poorly understood. A study on the effects of environmental factors of the freshwater oligochaete *Limnodrilus hoffmeisteri* in a South Korean stream is presented by Kong et al. Using multivariate analyses and a machine learning algorithm based on a nationwide scale database, the authors prove that water depth, velocity, and altitude are highly important environmental factors influencing the distribution of *Limnodrilus hoffmeisteri*, a species that is a recommended candidate to mitigate organic-enriched freshwater ecosystem.

Despite the awareness of the impact of climate change on rare endemic species, research on diversity, distribution, and conservation of endemic species remains limited, especially with respect to endemic macroinvertebrates on nationwide scales. Among aquatic insects, in particular, species in Ephemeroptera, Plecoptera, and Trichoptera (EPT) exhibit sensitive responses to physical environmental factors at broad scales, in addition to water quality factors at small scales. The diversity of EPT represents one of the most important biological indices for evaluating the status of freshwater habitats. Bae and Park identified the biogeographical and environmental factors affecting the biodiversity of endemic EPT in South Korea. They investigated the distribution pattern of endemic EPT species by applying non-metric multidimensional scaling (NMS) using the Bray–Curtis distance as the dissimilarity measure, predicting occurrence probability of endemic species using a random forest (RF) model with 39 environmental factors as independent variables. Geological and meteorological factors are identified as the main factors influencing species distribution. The results of this study support the need for an environmental management policy to regulate deforestation and conserve biodiversity, including endemic species.

Despite being described more than 20 years ago, evolutionary toxicology has only recently been proposed for ecotoxicological assessment to estimate long-term extinction risk, multigenerational effects, and effects of substances (or mixtures of substances) at sub-lethal environmental concentrations. The integration of next-generation sequencing (NGS) approaches has allowed for the identification of mechanisms and processes of adaptation to toxic substances. In the last paper in this volume, Rusconi et al. provide a review of current trends in this specific discipline, with a focus on population genetics and genomics approaches. They provide several examples indicating that evolutionary change may occur more rapidly in our lifetime. They also evidence that human activities are not only affecting the demography and the ecology of wild species, but also their evolutionary trajectory. They demonstrate the potential usefulness of predictive simulation and Bayesian techniques, also providing guidelines for a future implementation of evolutionary perspective into ecological risk assessment.

Marina Manca, Roberta Piscia and Piero Guilizzoni

Special Issue Editors

Article

Ammonium Transformation in 14 Lakes along a Trophic Gradient

Barbara Leoni *, Martina Patelli, Valentina Soler and Veronica Nava

Department of Earth and Environmental Sciences, University of Milano-Bicocca, Piazza Della Scienza 1, 20126 Milano, Italy; m.patelli3@campus.unimib.it (M.P.); valentina.soler@unimib.it (V.S.); v.nava15@campus.unimib.it (V.N.)

* Correspondence: barbara.leoni@unimib.it; Tel.: +39-02-6448-2712

Received: 3 February 2018; Accepted: 1 March 2018; Published: 3 March 2018

Abstract: Ammonia is a widespread pollutant in aquatic ecosystems originating directly and indirectly from human activities, which can strongly affect the structure and functioning of the aquatic foodweb. The biological oxidation of NH_4^+ to nitrite, and then nitrate is a key part of the complex nitrogen cycle and a fundamental process in aquatic environments, having a profound influence on ecosystem stability and functionality. Environmental studies have shown that our current knowledge of physical and chemical factors that control this process and the abundance and function of involved microorganisms are not entirely understood. In this paper, the efficiency and the transformation velocity of ammonium into oxidised compounds in 14 south-alpine lakes in northern Italy, with a similar origin, but different trophic levels, are compared with lab-scale experimentations (20 °C, dark, oxygen saturation) that are performed in artificial microcosms (4 L). The water samples were collected in different months to highlight the possible effect of seasonality on the development of the ammonium oxidation process. In four-liter microcosms, concentrations were increased by 1 mg/L NH_4^+ and the process of ammonium oxidation was constantly monitored. The time elapsed for the decrease of 25% and 95% of the initial ion ammonium concentration and the rate for that ammonium oxidation were evaluated. Principal Component Analysis and General Linear Model, performed on 56 observations and several chemical and physical parameters, highlighted the important roles of total phosphorus and nitrogen concentrations on the commencement of the oxidation process. Meanwhile, the natural concentration of ammonium influenced the rate of nitrification ($\mu g\ NH_4^+/L$ day). Seasonality did not seem to significantly affect the ammonium transformation. The results highlight the different vulnerabilities of lakes with different trophic statuses.

Keywords: lab-microcosms; ammonium impact; nitrification; trophic degree; lake vulnerability

1. Introduction

Total ammonia (TAN), in particular the unionized compound, is one of the major environmental pollutants in freshwater aquatic systems that is physiologically harmful to aquatic organisms and affects ecosystem functionality [1,2]. However, the threshold of ammonia toxicity varies widely, as there are sensitive and insensitive species. Nitrogen pollution in water has become a serious global environmental problem. It causes water eutrophication, stimulating the growth of dinoflagellates and Cyanobacteria and influencing phytoplankton blooms, and represents a potential hazard to human health [3–6].

In aquatic environments, total ammonia exists in two chemical forms, unionized ammonia (NH_3) and ionized ammonium (NH_4^+) [7], with different percentage depending on pH. In general, in water at 8.0 pH and 20 °C, only about 10% of the total ammonia is present as the more toxic form, ammonia (NH_3). Since 90% is present as ammonium (NH_4^+), it is preferable to use the term ammonium to refer to this type of pollution in natural water ([3,8] and the references therein).

Nowadays, there is increasing attention and a significant number of studies that are focusing on nitrogen to gain more knowledge about the factors that are influencing its different transformation pathways. In freshwater ecosystems, under anoxic conditions, the anaerobic oxidation of ammonium can occur, which is called anammox reaction: $NH_4^+ + NO_2^- \rightarrow N_2$ [9,10]. On the other hand, in aerobic conditions, the biological oxidation of ammonium to nitrite and then nitrate (nitrification) is a two-step process involving different taxa of chemolithotrophic organisms: archaeal and bacterial ammonia oxidizers (AOA, AOB), which obtain their energy from the oxidation of ammonia to nitrite, and nitrite oxidizing bacteria (NOB), which strictly depend on ammonia oxidizers and complete the oxidation to nitrate [11,12]. This process is a key part of the complex nitrogen cycle and a fundamental process in aquatic environments, having a profound influence on ecosystem stability [13,14].

Total ammonia can enter water bodies from natural sources, such as the end product of animal protein catabolism, and/or anthropogenic sources, such as atmospheric deposition, sewage effluents, industrial wastes, agricultural run-off, and the decomposition of biological wastes ([15–18] and references therein). In the last decades, anthropogenic activity has increased the inputs of nitrogen affecting water quality and aquatic biodiversity, with also considerable socioeconomic consequences [19–22]. Thus, environmental factors influencing ammonium oxidation in freshwater systems have received considerable attention. Previous studies have highlighted that the nitrification rate in estuaries and in rivers depends on the activities of nitrifying bacteria and is affected by environmental parameters such as temperature, light, and pH values, as well as oxygen, nitrogen, organic carbon, and sulphide concentrations [13,23–25]. Despite recent advances, measurements of rates and controls of nitrification are relatively rare in lake ecosystems [26]. There is limited knowledge of the relationships among ammonium nitrification rate, lake trophic degree, and the associated microorganisms, which are closely related with the NH_4^+ removal efficiency and the self-purification capacity of lake ecosystems [27–29].

The transformation efficiency and velocity of ammonium into oxidized compounds in 14 lakes, located in the same geographic region, were compared with lab-scale experimentations performed in artificial aerobic microcosms. Our major goals were to determine nitrification rates in several lentic environments, which are characterized by different trophic levels (e.g., Total Phosphorus, Total Nitrogen) and natural content levels of ammonium in different seasons.

We conducted experiments to address the following questions that arose from previous studies: Are nitrification rates related to the lake trophic degree? How do lakes with different trophic degree react to an increased load of ammonium? Is the nitrification process influenced by seasonality? We hypothesized that the nitrification rate increased in productive lakes and in some seasons. A possible relationship between there parameters could lead to further questions regarding the different vulnerabilities in lakes of different trophic statuses.

2. Materials and Methods

2.1. Study Sites

Water samples were collected from fourteen Italian lakes: L. Candia, L. Orta, L. Mergozzo, L. Maggiore, L. Monate, L. Comabbio, L. Varese, L. Piano, L. Montorfano, L. Alserio, L. Segrino, L. Pusiano, L. Annone (W), and L. Olginate. These lakes are located in Northern Italy, included from Piemonte region (45°49′ N, 8°24′ E) and the western part of Lombardia region (45°47′ N, 9°25′ E) (Figure 1).

The lakes have different morphometric and chemical characteristics. The lakes were classified in relation to their different trophic status, following Organisation for Economic Co-operation and Development recommendations [30], which span from oligotrophic to highly eutrophic. For more information about the 14 studied lakes, see Table 1.

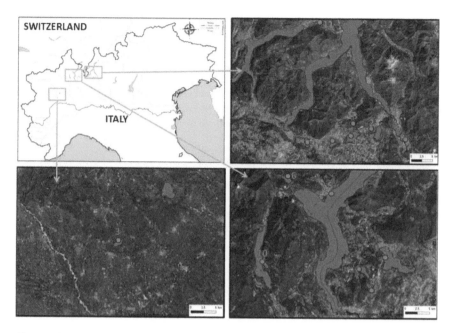

Figure 1. Locations of the 14 sampling south-alpine lakes in Northern Italy. The numbers in the map are referred to lakes list in Table 1 (from d-maps.com modified).

Table 1. Main morphometric and chemico-physical characteristics of 14 south-alpine lakes [31–35].

Lake	Area (km^2)	Volume (m^3 × 10^6)	Depth$_{max}$ (m)	Depth$_{mean}$ (m)	Trophic Status
1-Alserio	1.23	6.55	8.1	5.3	Hypereutrophic
2-Annone (W)	1.75	7.60	10.0	4.5	Eutrophic
3-Candia	1.49	8.10	7.7	3.8	Meso-eutrophic
4-Comabbio	3.58	16.40	7.7	4.6	Hypereutrophic
5-Maggiore	213	37500	370	176	Oligotrophic
6-Mergozzo	1.83	83.00	73.0	45.4	Oligotrophic
7-Monate	2.51	45.00	34.0	14.4	Oligotrophic
8-Montorfano	1.90	1.90	6.8	4.0	Meso-eutrophic
9-Olginate	0.58	7.00	17.0	8.0	Meso-eutrophic
10-Orta	18.10	1286	143	70.9	Oligotrophic
11-Piano	0.63	4.03	12.5	5.1	Eutrophic
12-Pusiano	4.93	69.00	24.3	14.0	Eutrophic
13-Segrino	0.38	1.20	9.0	4.6	Oligotrophic
14-Varese	14.90	162	26.0	9.9	Hypereutrophic

2.2. Sampling Method and Experimental Procedure

For each lake, the water samples were collected in different months to highlight the possible variations in the ammonium transformation in relation to seasonality, totaling 74 samples. On water samples of Lakes Candia and Piano, the experiment was performed only once during the study, because, for logistic problems, we could not collect their water in different months. Each lake sample was collected by Niskin's bottle near the littoral zone (3 m depth), in open water without vegetation, and quickly transferred to the laboratory. Before the experimental procedure, we measured the main chemical and physical parameters, characterizing lacustrine water quality. Temperature and pH were detected in situ with a portable underwater multiparameter probe (WTW multi3432). Total phosphorus, soluble reactive phosphorus, total nitrogen, ammonium nitrogen, and alkalinity were analyzed using standard methods [36,37]. Nitrate nitrogen anions were measured using the Ion Chromatography (Thermo Scientific™ Dionex™, Waltham, MA, USA).

For each lake, the water samples were placed into clean 4 L polycarbonate microcosms (tanks) and amended with one mg/L of ammonium, as NH_4Cl (10 μM). The microcosms were kept in the dark at 20 °C and continually shaken. The oxygen saturation was maintained between 80% and 100%. For all the sampled lakes, excluding L. Candia and L. Piano (not watertight seal of microcosms), experiment replicates have been performed in order to verify the text repeatability.

The ammonium consumption was monitored at high frequency, by colorimetric spectrophotometry, until its complete depletion.

2.3. Data Analyses

To analyze and discuss the experimental results, three parameters have been evaluated: (a) D25 and (b) D95, as time elapsed to observe an ammonium decrease of 25% and 95% of the initial concentration in amended microcosm; (c) OxRate, amount of ammonium oxidised in each day (μg NH_4^+/L day) to reduce the ammonium concentration from 25% to 95% of initial concentration in amended microcosms. To avoid redundancy among dependent variables, the relations between these parameters have been tested by Spearman's rank correlation coefficient.

For statistical analyses, two parameters were selected as response variables, D25 and OxRate, and to avoid collinearity among predictor, two principal component analyses (PCA) were performed in order to reduce dimensions and select a smaller set of variables [38]. The chemico-physical included parameters, measured before the experiment on lake water samples, were: total phosphorus (TP), soluble reactive phosphorus (SRP), total nitrogen (TP), nitrate nitrogen (NO_3^-), ammonium nitrogen (NH_4^+), temperature (TEMP), alkalinity (ALK), and pH. All data were centered (mean value = 0) and scaled (variance = 1) to allow comparison among parameters [39]. Only six predictors were selected from the initial set of 8 to perform the following analyses. We used general linear models (GLMs), a statistical procedure similar to an analysis of variance used to estimate effect size of different factors on a variable of interest [40,41], to analyze the effect of trophic degrees and seasonality on the transformation velocity of ammonia into oxidised compounds. One model had as response variable D25 and the other one had OxRate. We first built a full model including all independent variables that may affect the dependent variable under scrutiny: TP, TN, ALK, TEMP, pH, NH_4^+, and the possible interactions. Finally, we removed all of the non-significant predictors in two-steps to obtain a final model [42]. The assumptions for general linear models were checked by inspection of diagnostic plots and applying Shapiro–Wilks tests [43]. Interactions were excluded before the relevant main effects. Statistical analyses and figures were produced using different packages (base packages and "ggplot2", "corrplot", "factoextra") in R 3.4.1. [44–47].

3. Results

3.1. Lakes Characteristics

The mean values (±Standard Error of the Mean) of chemico-physical parameters measured for each lake during every sampling activity are reported in Table 2. Among the 14 lakes that were analyzed, total phosphorus (TP) ranged from 3.0 to 144.0 μg/L, with a mean value of 21.7 ± 2.3 μg/L (±SEM) and soluble reactive phosphorus (SRP) ranged from 1.0 to 99.0 μg/L, with a mean value of 9.8 ± 1.9 μg/L. Total nitrogen (TN) concentrations were between 90.0 and 4218.5 μg/L, with a mean value of 1173.1 ± 93.1 μg/L. Nitrate nitrogen (NO_3^-) and ammonium nitrogen (NH_4^+) were in the range of 0–3770 μg/L—with a mean of 593.0 ± 83.2 μg/L, and 0–775 μg/L—with a mean value of 125.9 ± 24.0 μg/L, respectively. Temperature showed a mean value of 14.3 ± 0.9 °C, ranging between 5.6 °C and 27.3 °C. Alkalinity ranged from 0.3 to 4.3, with a mean value of 2.0 ± 0.1 meq/L. Finally, pH had a mean value of 7.9 ± 0.06 and ranged from 6.5 to 8.9.

Table 2. Mean value ±SEM of the chemico-physical parameters of the 14 lakes, calculated on measurements performed in different months. Note that for Lake Candia and Piano the values refer to a single month. n: number of experiments performed for each lake.

Lake	n	TP (µg/L)	SRP (µg/L)	TN (µg/L)	NO_3^- (µg/L)	NH_4^+ (µg/L)	Temperature (°C)	Alkalinity (meq/L)	pH (Units)
1-Alserio	5	31.9 ± 2.4	17.5 ± 3.4	2930 ± 333	1680 ± 543	320 ± 137	13.7 ± 2.8	3.8 ± 0.2	8.0 ± 0.1
2-Annone (W)	4	27.1 ± 2.2	8.7 ± 0.5	870 ± 79	171 ± 82	87 ± 43	15.1 ± 4.0	2.9 ± 0.1	8.3 ± 0.1
3-Candia	1	18.0	2.0	1060	170	17	27	1.18	7.5
4-Comabbio	4	30.5 ± 5.5	16.2 ± 3.6	860 ± 105	158 ± 88	89 ± 75	15.2 ± 4.0	2.0 ± 0.1	8.4 ± 0.1
5-Maggiore	4	7.6 ± 0.7	3.5 ± 0.6	885 ± 53	648 ± 61	10 ± 5	14.4 ± 3.5	0.9 ± 0.2	7.9 ± 0.2
6-Mergozzo	4	5.0 ± 0.7	2.5 ± 0.3	851 ± 64	595 ± 36	15 ± 10	14.8 ± 3.8	0.3 ± 0.1	7.3 ± 0.3
7-Monate	3	8.0 ± 1.3	2.0 ± 0.6	459 ± 13	158 ± 27	16 ± 6	13.6 ± 4.0	1.1 ± 0.2	7.9 ± 0.5
8-Montorfano	7	18.3 ± 2.5	6.8 ± 0.6	936 ± 84	129 ± 50	245 ± 94	16.3 ± 2.9	1.9 ± 0.1	8.3 ± 0.1
9-Olginate	4	18.0 ± 1.9	5.7 ± 1.8	942 ± 60	648 ± 79	39 ± 7	13.5 ± 3.5	1.4 ± 0.3	8.2 ± 0.3
10-Orta	4	7.2 ± 1.4	3.2 ± 0.5	1537 ± 19	1313 ± 24	16 ± 7	14.7 ± 3.7	0.3 ± 0.1	7.2 ± 0.3
11-Piano	1	63.0	8.0	2069	1424	677	11	3.8	8.4
12-Pusiano	4	26.0 ± 4.6	3.6 ± 0.5	1313 ± 256	667 ± 189	133 ± 89	14.8 ± 4.4	2.6 ± 0.2	8.1 ± 0.2
13-Segrino	7	10.7 ± 0.7	4.0 ± 0.5	1311 ± 195	751 ± 198	85 ± 16	11.4 ± 2.5	2.4 ± 0.1	7.9 ± 0.1
14-Varese	5	50.4 ± 19.1	38.2 ± 18.3	901 ± 104	220 ± 71	117 ± 88	12.7 ± 3.1	2.5 ± 0.1	7.9 ± 0.1

3.2. Ammonium Oxidation

The trend of ammonium oxidation in the various lakes and in water samples collected in different months are shown in Figure 2. The oxidation process, after the addition of 1 mg/L of NH_4^+, started after several days in all microcosms with different timing depending on the lake. The minimum was observed in microcosm of L. Annone sampled in June (Figure 2 (2-Annone)) while the maximum in L. Mergozzo sampled in October (Figure 2 (6-Mergozzo)). It was not possible to observe a clear relationship between sampling month and the time elapsed for the ammonium oxidation.

For all of the 74 water samples examined, D25 ranged from 6.5 to 34.8 days, with a mean value of 15.1 ± 0.7 days (±SEM). The various lakes showed a difference in this parameter, with a mean value of 8.9 ± 0.5 days for L. Alserio, 12.2 ± 2.1 days for L. Annone, 7.5 days for L. Candia, 15.5 ± 2.2 days for L. Comabbio, 14.5 ± 1.9 days for L. Maggiore, 24.8 ± 3.9 days for L. Mergozzo, 19.7 ± 2.5 days for L. Monate, 16.7 ± 2.3 days for L. Montorfano, 10.6 ± 0.7 days for L. Olginate, 18.0 ± 2.2 days for L. Orta, 16.5 days for L. Piano, 11.9 ± 1.7 days for L. Pusiano, 17.7 ± 4.3 days for L. Segrino, 13.2 ± 2.1 days for L. Varese (Figure 3). The lakes that showed the lower velocity of ammonium oxidation were L. Mergozzo, L. Monate, L. Orta, L. Segrino; meanwhile, L. Candia, L. Alserio, and L. Olginate displayed the higher velocity of ammonium oxidation. A huge variation in D25 value was highlighted in the various months for L. Mergozzo, for which we detected a difference between the maximum and the minimum value of 19.3 days. However, the overall D25 mean value of the fourteen lakes did not display a large variation among the different seasons. This parameter had a mean value of 15.2 ± 1.4 days in spring, 15.2 ± 1.6 days in summer, 16.7 ± 1.9 days in autumn, and 13.2 ± 1.0 days in winter.

D95 showed a mean value among the lakes of 17.9 ± 0.8 days. The various lakes showed a difference in this parameter, with a mean value of 12.8 ± 1.3 days for L. Alserio, 14.6 ± 2.2 days for L. Annone, 13.8 days for L. Candia, 17.7 ± 2.2 days for L. Comabbio, 16.8 ± 2.2 days for L. Maggiore, 27.8 ± 4.1 days for L. Mergozzo, 21.7 ± 2.2 days for L. Monate, 19.7 ± 2.4 days for L. Montorfano, 13.7 ± 0.5 days for L. Olginate, 20.6 ± 2.8 days for L. Orta, 20.5 days for L. Piano, 14.3 ± 1.7 days for L. Pusiano, 20.4 ± 2.2 days for L. Segrino, 16.5 ± 1.7 days for L. Varese (Figure 3). The lakes that showed the higher value of D95 were L. Mergozzo, L. Monate; instead L. Alserio, and L. Olginate displayed the lower values, similar to that detected for D25. The results were similar to that detected for D25 and a Spearman's correlation analysis highlighted a strong relationship ($\rho = 0.98$, $p < 0.001$) between D25 and D95.

The difference in days elapsed from the D95 to D25 ranged from 1.3 (L. Orta in February) to 5.0 days (L. Segrino in December) with a mean value of 3.9 ± 0.2 days.

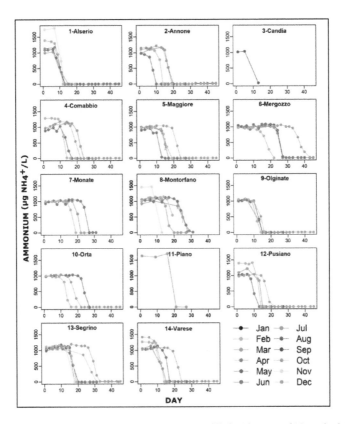

Figure 2. Ammonium oxidation trends in lab-microcosms filled with water of 14 south-alpine lakes sampled in different months and amended with 1 mg /L of NH_4^+. Results are expressed as mean ±SEM for duplicate tests. Note that the SEM bars in most of cases showed low values.

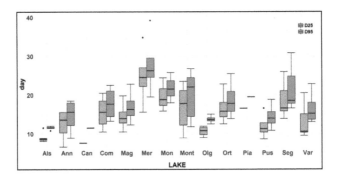

Figure 3. Box plot of the number of days elapsed from the experiment start until an ammonia concentration decrease of 25% (D25) and 95% (D95) of the initial concentration in the 14 studied lakes. Note that for samples of Lakes Candia and Piano the experiment was performed only once during the study. Box plot statistics: the lower and upper hinges correspond to the first and third quartiles. The upper (lower) whisker extends from the hinge to the largest (smallest) value no further than $1.5 \times InterQuartileRange$ from the hinge. Data beyond the end of the whiskers are outlying points and are plotted individually.

The rate of oxidation (OxRate), referred to the time elapsed to measure an ammonium decrease from 25% to 95%, showed different values from 90 to 400 µg NH₄⁺/L day, respectively, in L. Mergozzo, in October, and L. Varese, in June. The mean value of the 14 lakes was equal to 210 ± 9.7 µg NH₄⁺/L day (Figure 4).

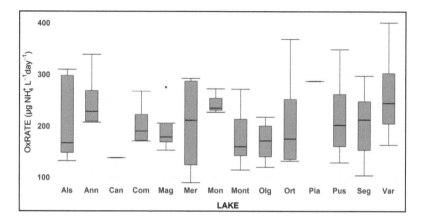

Figure 4. Box plot of the OxRate, amount of ammonium oxidised in each day (µg NH₄⁺/L day) to reduce the ammonium concentration from 25% to 95% of the initial concentration in amended microcosms. Note that for samples of Lakes Candia and Piano the experiment was performed only once during the study. See Figure 3 for box plot statistics.

3.3. Relationship between Ammonium Oxidation and Chemico-Physical Parameters

The loadings plots of the Principal Component Analyses performed on the eight chemico-physical parameters and D25 and OxRate are presented in Figure 5.

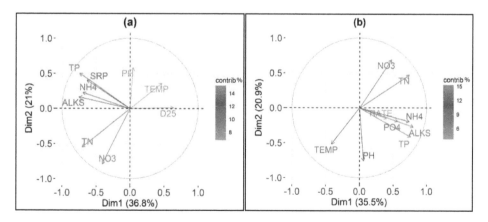

Figure 5. Loadings plots of the first two components of the principal component analysis (PCA) with different colors according to the parameters contributions ("contrib %") performed on 8 chemico-physical parameters and (**a**) number of days elapsed from the beginning of the experimental procedure until an ammonium concentration decrease equal to the 25% (D25); and, (**b**) oxidation rate of ammonium (OxRate). See "Materials and Methods" for abbreviations.

The first two components of the PCA with the D25 values explained 57.8% of the total variance (PC1 36.8%; PC2 21%—Figure 5a); while, the first two components of the PCA with the OxRate values explained 56.4% of the total variance (PC1 35.5%; PC2 20.9%—Figure 5b). Total phosphorus and soluble reactive phosphorus were highly correlated. There was also a strong correlation between total nitrogen and nitrate nitrogen. Consequently, soluble reactive phosphorus and nitrate nitrogen were removed from the following analyses to avoid collinearity.

We used general linear models (GLM) in order to analyze the effect of chemical and physical parameters, trophic condition (as indicated by TP and TN), and seasonality on the oxidation velocity of ammonium. General Linear Models (GLMs) were performed with D25 and OxRate as response variables. The full models were built using six parameters that included non-significant terms which, following a two-step process, were excluded from the models (see "Materials and Methods" section). The original in-lake values for temperature, alkalinity, ammonium nitrogen, pH and sampling months seemed not to be a significant predictor for D25. The final GLM highlighted total phosphorus and total nitrogen as significant predictors of time elapsed to oxidase the 25% of initial ammonium (Table 3a). The D25 model was significant ($F_{(54)} = 11.36$, $p < 0.001$).

Only ion ammonium initial concentration (NH_4^+) seemed to be a significant predictor for OxRate (Table 3b). The overall OxRate model was significant ($F_{(54)} = 11.71$, $p < 0.01$).

Table 3. Final General Linear Model and parameter estimates of (**a**) the response variable D25 and (**b**) the response variable OxRate. Number of tested data was 56 in each model and the months included ranged from February to December.

Model	Response Variable	Predictor	t	p
(a)	D25	TP	−3.133	0.003 **
		TN	−2.974	0.004 **
(b)	OxRate	NH_4^+	3.422	0.002 **

Note: ** $p < 0.01$.

4. Discussion

During the last decades, human population growth and anthropic activities have substantially altered the nitrogen cycle, increasing the availability and the mobility of this element. The enhanced concentrations of inorganic nitrogenous compounds (NH_4^+, NO_2^-, NO_3^-) in ground and surface waters affect many aquatic organisms and contribute to the degradation of freshwater, estuarine, and coastal marine ecosystems [48,49]. In particular, intensive animal farming and the use of fertilizer have significantly increased ammonium loading. Unionized ammonia is very toxic to aquatic communities, in particular fish, causing, for example, damage to the gill epithelium, acidosis, ATP depletion in the basilar region of the brain, disruption of osmoregulatory activity, etc. [48]. The ionized form of ammonia, ammonium (NH_4^+), has minor toxicity compared to NH_3 [48]. On the basis of acute and chronic toxicity data, water quality criteria were proposed to protect sensitive aquatic animals, which was equal to 17 mg TAN/L for short-term exposure and 1.9 mg TAN/L for long-term exposure (pH = 7; T = 20 °C) [48,50]. In the present study, the natural concentrations of total ammonia in the 14 studied lakes varied widely, with values up to 0.77 mg/L (43 μM). This value is below the toxicity threshold but could represent a problem of concern in the future, given the reported increase of these compounds in aquatic systems [51,52]. High ammonium loads can also enhance the development of primary producers. Historically, phosphorus has been considered the primary limiting nutrient for phytoplanktonic organisms; however, recent studies have highlighted the importance of nitrogen in determining both the proliferation and the toxicity of non-diazotrophic Cyanobacteria [48,53,54]. Indeed, ammonium appears to favor cyanobacterial dominance in lakes, most likely because of their superior uptake kinetics [3]. At the same time, when NH_4^+ concentrations are sufficiently high, ammonium may suppress the overall growth of phytoplanktonic organisms [3,55]. Significant inputs of NH_4^+ can also contribute to the acidification process since nitrification produces hydrogen ions [48], and consequently contributing to the loss of aquatic plants [56,57].

In this context, more knowledge about the transformation process of ammonium is needed, particularly concerning the vulnerability of different lakes to ammonium changes. The goal of the study was to determine the 'ability' of lakes with different trophic levels and natural content levels of TAN to overcome high concentrations of ammonium. Our study highlighted a significant difference in the ammonium oxidation rate in water of 14 lakes amended with 1 mg NH_4^+/L. Previous studies have highlighted that a high ammonia concentration may result in ammonia oxidation inhibition; however, to observe this effect there must be a recorded ammonia concentration in the range of 10–150 mg/L [58–60]. We are confident that an ammonium decrease in the microcosms was linked to nitrification processes, as the experiments were conducted in aerobic conditions and the anammox process could not occur.

Using a lab-scale approach, we found that the start of the nitrification process varied as a function of the lake's trophic status. Indeed, lakes that are characterized by a higher trophic degree seemed to present a faster ammonium transformation, as the nitrification process started earlier when compared to the samples from the oligotrophic lakes. The multivariate regression supported this statement, as the number of days elapsed from the experiment start until a decrease of ammonium concentration equal to 25% was inversely related to total phosphorus and total nitrogen contents. Total phosphorus and nitrogen represent the limiting nutrients in freshwater ecosystems; thus, their concentrations can be considered representative of the lake trophic status [61]. Previous studies have reported the importance of these two nutrients in determining the abundance of archaea and bacteria that are responsible for the ammonium nitrification process [62]. The pH did not seem to affect the D25 parameter, as no relationship was highlighted both in PCA and in GLM. This result appeared inconsistent with those found by different studies, since the effect of pH on nitrification has been widely shown. The literature reported that complete inhibition of nitrification could occur at a pH lower than 6.35 and higher than 8.95; meanwhile, from 6.45 to 8.95, there was no effect and complete nitrification occurred [63]. In the present work, the initial pH values were included in this range, which were always more than 6.5 and less than 8.9, and no nitrification inhibition could be highlighted, as we did not observe a postponed nitrification start in relation to the lower values of pH. Similar observations could be reported for alkalinity, which has been widely reported to affect the nitrification process [64,65]. Alkalinity is an important parameter, as it influences pH and is a measure of the inorganic carbon available for nitrifiers; a lack of alkalinity will stop nitrification. Alkalinity is consumed at 7.14 g as $CaCO_3$ per gram of NH_4^+ oxidized to NO_2^- and, consequently, it decreases during the nitrification process. Optimal values for nitrification range from 75 to 150 mg $CaCO_3$/L [66,67]. Our statistical analyses did not show a significant influence of the initial alkalinity (ranging from 13 to 216 mg $CaCO_3$/L) on the number of days elapsed from the experiment start until a decrease of ammonium concentration equal to 25% was achieved.

We also evaluated the effect of seasonality on the nitrification process, based on the lake temperature detected when the water samples were collected. Our results did not show any effects of the lake in-situ temperature on the commencement of oxidation in the lab-microcosms, even though the samples collected in different months provided lake water temperatures ranging from 5.6 to 27.3 °C. Indeed, the temperature affects the growth rate of both types of bacteria involved in the nitrification process, with ammonia oxidizers having superior growth rates at high temperatures when compared to nitrite oxidizers [63]. We hypothesized that in lab-microcosms, which were constantly maintained at 20 °C, the commencement of oxidation was affected by the bacteria acclimation lag time, as Zhu and Chen [54] reported that nitrifying bacteria can adapt to a wide range of temperatures if acclimated slowly. However, it has to be highlighted that using temperature as an indicator of seasonality could restrict the analysis of this predictor factor. In effect, several lakes showed a higher rate of oxidation at the start (February–March) compared to the end (October–November) of the growing season.

Although a clear seasonal pattern could not be highlighted overall, some lakes showed a difference in the D25 value among the experiments. However, these differences did not appear to be related to the month, since for some lakes (e.g., L. Maggiore, L. Mergozzo, L. Pusiano, L. Segrino, L. Varese) the higher value of D25 was recorded in autumn, while for others (e.g., L. Monate, L. Montorfano, L. Orta), the higher value was recorded in summer. Furthermore, the interaction between seasonality and nutrient contents (in GLM) was not selected among the predictive parameters of the decrease in ammonium concentrations and of the oxidation rate in microcosms. Future research with a field-scale approach should help to clarify the role of seasonality in the nitrification process.

Our research highlighted an important role of the initial acclimation in the determination of the time needed to oxidize all ammonium, as the range of the D25 parameter was quite large compared to the range of days elapsed from D25 to D95. Once it began, the ammonium oxidation rate was significantly and positively affected by the natural content of ammonium in the lakes, as highlighted by the second GLM model [12,62,68].

In conclusion, the time required to oxidize a 1 mg/L ammonium addition in microcosms filled with water of 14 south-alpine lakes showed a remarkable range. Lake trophic status and nutrient concentrations have been shown to play important roles in determining the time elapsed before the beginning of nitrification processes, while the natural ammonium concentration in lakes significantly influences the nitrification rate of additional ammonium inputs.

Acknowledgments: This work was supported by the University of Milano-Bicocca (FA). We are grateful to Letizia Garibaldi, Alba Varallo and Angela Ferrauto. We thank the anonymous reviewers for their comments, which helped to improve this article.

Author Contributions: B.L. conceived the study, performed lab experiments, discussed results, wrote the text; M.P. collected samples, performed lab analyses and discussed results; V.S. collected samples and performed lab analyses; V.N. performed statistical analyses, discussed results and wrote the text.

Conflicts of Interest: The authors declare no conflict of interest.

References

1. Benli, A.Ç.K.; Köksal, G.; Özkul, A. Sublethal ammonia exposure of Nile tilapia (*Oreochromis niloticus* L.): Effects on gill, liver and kidney histology. *Chemosphere* **2008**, *72*, 1355–1358. [CrossRef] [PubMed]
2. Li, M.; Yu, N.; Qin, J.G.; Li, E.; Du, Z.; Chen, L. Effects of ammonia stress, dietary linseed oil and Edwardsiella ictaluri challenge on juvenile darkbarbel catfish *Pelteobagrus vachelli*. *Fish Shellfish Immunol.* **2014**, *38*, 158–165. [CrossRef] [PubMed]
3. Collos, Y.; Harrison, P.J. Acclimation and toxicity of high ammonium concentrations to unicellular algae. *Mar. Pollut. Bull.* **2014**, *80*, 8–23. [CrossRef] [PubMed]
4. Zekker, I.; Rikmann, E.; Tenno, T.; Kroon, K.; Vabamäe, P.; Salo, E.; Tenno, T.; Loorits, L.; Dc Rubin, S.S.C.; Vlaeminck, S.E. Deammonification process start-up after enrichment of anammox microorganisms from reject water in a moving-bed biofilm reactor. *Environ. Technol. (United Kingdom)* **2013**, *34*, 3095–3101. [CrossRef] [PubMed]
5. Zekker, I.; Rikmann, E.; Tenno, T.; Vabamäe, P.; Kroon, K.; Loorits, L.; Saluste, A.; Tenno, T. Effect of concentration on anammox nitrogen removal rate in a moving bed biofilm reactor. *Environ. Technol.* **2012**, *33*, 2263–2271. [CrossRef] [PubMed]
6. Leoni, B.; Garibaldi, L. Population dynamics of *Chaoborus flavicans* and *Daphnia* spp.: Effects on a zooplankton community in a volcanic eutrophic lake with naturally high metal concentrations (L. Monticchio Grande, Southern Italy). *J. Limnol.* **2009**, *68*, 37–45. [CrossRef]
7. Wajsbrot, N.; Gasith, A.; Diamant, A.; Popper, D.M. Chronic toxicity of ammonia to juvenile gilthead seabream *Sparus aurata* and related histopathological effects. *J. Fish Biol.* **1993**, *42*, 321–328. [CrossRef]
8. Hargreaves, J.A.; Tucker, C.S. *Managing Ammonia in Fish Ponds*; SRAC Publication: Stoneville, MS, USA, 2004.
9. Brandes, J.A.; Devol, A.H.; Deutsch, C. New developments in the marine nitrogen cycle. *Chem. Rev.* **2007**, *107*, 577–589. [CrossRef] [PubMed]

10. Zekker, I.; Rikmann, E.; Kroon, K.; Mandel, A.; Mihkelson, J. Ameliorating nitrite inhibition in a low-temperature nitritation–anammox MBBR using bacterial intermediate nitric oxide. *Int. J. Environ. Sci. Technol.* **2017**, *14*, 2343–2356. [CrossRef]

11. Hayden, C.J.; Beman, J.M. High abundances of potentially active ammonia-oxidizing bacteria and archaea in oligotrophic, high-altitude lakes of the Sierra Nevada, California, USA. *PLoS ONE* **2014**, *9*, e111560. [CrossRef] [PubMed]

12. Mukherjee, M.; Ray, A.; Post, A.F.; McKay, R.M.; Bullerjahn, G.S. Identification, enumeration and diversity of nitrifying planktonic archaea and bacteria in trophic end members of the Laurentian Great Lakes. *J. Great Lakes Res.* **2016**, *1*, 39–49. [CrossRef]

13. Bernhard, A.E.; Tucker, J.; Giblin, A.E.; Stahl, D.A. Functionally distinct communities of ammonia-oxidizing bacteria along an estuarine salinity gradient. *Environ. Microbiol.* **2007**, *9*, 1439–1447. [CrossRef] [PubMed]

14. Wang, C.; Liu, J.; Wang, Z.; Pei, Y. Nitrification in lake sediment with addition of drinking water treatment residuals. *Water Res.* **2014**, *56*, 234–245. [CrossRef] [PubMed]

15. Randall, D.J.; Tsui, T.K.N. Ammonia toxicity in fish. *Mar. Pollut. Bull.* **2002**, *45*, 17–23. [CrossRef]

16. Sinha, A.K.; Liew, H.J.; Diricx, M.; Blust, R.; De Boeck, G. The interactive effects of ammonia exposure, nutritional status and exercise on metabolic and physiological responses in gold fish (*Carassius auratus* L.). *Aquat. Toxicol.* **2012**, *109*, 33–46. [CrossRef] [PubMed]

17. Cheng, C.H.; Yang, F.F.; Ling, R.Z.; Liao, S.A.; Miao, Y.T.; Ye, C.X.; Wang, A.L. Effects of ammonia exposure on apoptosis, oxidative stress and immune response in pufferfish (*Takifugu obscurus*). *Aquat. Toxicol.* **2015**, *164*, 61–71. [CrossRef] [PubMed]

18. Rikmann, E.; Zekker, I.; Tenno, T.; Saluste, A.; Tenno, T. Inoculum-free start-up of biofilm- and sludge-based deammonification systems in pilot scale. *Int. J. Environ. Sci. Technol.* **2018**, *15*, 133–148. [CrossRef]

19. Pauer, J.J.; Auer, M.T. Nitrification in the water column and sediment of a hypereutrophic lake and adjoining river system. *Water Res.* **2000**, *34*, 1247–1254. [CrossRef]

20. Beaulieu, J.J.; Tank, J.L.; Hamilton, S.K.; Wollheim, W.M.; Hall, R.O.; Mulholland, P.J.; Peterson, B.J.; Ashkenas, L.R.; Cooper, L.W.; Dahm, C.N.; et al. Nitrous oxide emission from denitrification in stream and river networks. *Proc. Natl. Acad. Sci. USA* **2011**, *108*, 214–219. [CrossRef] [PubMed]

21. Zhao, X.; Wei, Z.; Zhao, Y.; Xi, B.; Wang, X.; Zhao, T.; Zhang, X.; Wei, Y. Environmental factors influencing the distribution of ammonifying and denitrifying bacteria and water qualities in 10 lakes and reservoirs of the Northeast, China. *Microb. Biotechnol.* **2015**, *8*, 541–548. [CrossRef] [PubMed]

22. Smith, V.; Wood, S.; McBride, C.; Atalah, J.; Hamilton, D. Phosphorus and nitrogen loading restraints are essential for successful eutrophication control of Lake Rotorua, New Zealand. *Inland Waters* **2016**, *6*, 273–283. [CrossRef]

23. Lipschultz, F.; Wofsy, S.C.; Fox, L.E. The effects of light and nutrients on rates of ammonium transformation in a eutrophic river. *Mar. Chem.* **1985**, *16*, 329–341. [CrossRef]

24. Usui, T.; Koike, I.; Ogura, N. N_2O production, nitrification and denitrification in an estuarine sediment. *Estuar. Coast. Shelf Sci.* **2001**, *52*, 769–781. [CrossRef]

25. Strauss, E.A.; Richardson, W.B.; Bartsch, L.A.; Cavanaugh, J.C.; Bruesewitz, D.A.; Imker, H.; Heinz, J.A.; Soballe, D.M. Nitrification in the Upper Mississippi River: Patterns, controls, and contribution to the NO_3^- budget. *J. N. Am. Benthol. Soc.* **2004**, *23*, 1–14. [CrossRef]

26. Small, G.E.; Bullerjahn, G.S.; Sterner, R.W.; Beall, B.F.N.; Brovold, S.; Finlay, J.C.; McKay, R.M.L.; Mukherjee, M. Rates and controls of nitrification in a large oligotrophic lake. *Limnol. Oceanogr.* **2013**, *58*, 276–286. [CrossRef]

27. Leoni, B.; Nava, V.; Patelli, M. Relationship among climate variability, Cladocera phenology and pelagic food web in deep lakes in different trophic state. *Mar. Freshw. Res.* **2017**. accepted for publication.

28. Kowalczewska-Madura, K.; Godyn, R.; Dera, M. Spatial and seasonal changes of phosphorus internal loading in two lakes with different trophy. *Ecol. Eng.* **2015**, *74*, 187–195. [CrossRef]

29. Poste, A.E.; Muir, D.C.G.; Guildford, S.J.; Hecky, R.E. Bioaccumulation and biomagnification of mercury in African lakes: The importance of trophic status. *Sci. Total Environ.* **2015**, *506–507*, 126–136. [CrossRef] [PubMed]

30. Premazzi, G.; Chiaudani, G. *Ecological Quality of Surface Waters: Quality Assessment Schemes for European Community Lakes*; European Commission: Brussels, Luxembourg, 1992; Volume 14563.

31. Ambrosetti, W.; Barbanti, L. Physical limnology of Italian lakes. 1. Relationship between morphometry and heat content. *J. Limnol.* **2003**, *61*, 147–157. [CrossRef]

32. Leoni, B.; Morabito, G.; Rogora, M.; Pollastro, D.; Mosello, R.; Arisci, S.; Forasacco, E.; Garibaldi, L. Response of planktonic communities to calcium hydroxide addition in a hardwater eutrophic lake: Results from a mesocosm experiment. *Limnology* **2007**, *8*, 121–130. [CrossRef]

33. Riccardi, N.; Mangoni, M. Considerations on the biochemical composition of some freshwater zooplankton species. *J. Limnol.* **1999**, *58*, 58–65. [CrossRef]

34. Volta, P.; Jeppesen, E.; Leoni, B.; Campi, B.; Sala, P.; Garibaldi, L.; Lauridsen, T.L.; Winfield, I.J. Recent invasion by a non-native cyprinid (common bream *Abramis brama*) is followed by major changes in the ecological quality of a shallow lake in southern Europe. *Biol. Invasions* **2013**, *15*, 2065–2079. [CrossRef]

35. Rogora, M.; Mosello, R.; Kamburska, L.; Salmaso, N.; Cerasino, L.; Leoni, B.; Garibaldi, L.; Soler, V.; Lepori, F.; Colombo, L.; et al. Recent trends in chloride and sodium concentrations in the deep subalpine lakes (Northern Italy). *Environ. Sci. Pollut. Res.* **2015**, *22*, 19013–19026. [CrossRef] [PubMed]

36. American Public Health Association (APHA). *Standard Methods for the Examination of Water and Wastewater*; American Public Health Association: Washington, DC, USA, 1998; p. 552.

37. Marti, C.M.; Imberger, J.; Garibaldi, L.; Leoni, B. Using time scales to characterize phytoplankton assemblages in a deep subalpine lake during the thermal stratification period: Lake Iseo, Italy. *Water Resour. Res.* **2015**, *52*, 1762–1780. [CrossRef]

38. Van Der Maaten, L.J.P.; Postma, E.O.; Van Den Herik, H.J. Dimensionality Reduction: A Comparative Review. *J. Mach. Learn. Res.* **2009**, *10*, 1–41. [CrossRef]

39. Jollife, I.T.; Cadima, J. Principal component analysis: A review and recent developments. *Philos. Trans. A Math. Phys. Eng. Sci.* **2016**, *374*, 20150202. [CrossRef] [PubMed]

40. Madsen, H.; Thyregod, P. *Introduction to General and Generalized Linear Models*; CRC Press: Boca Raton, FL, USA, 2010; ISBN 9781420091557.

41. Leoni, B.; Marti, C.L.; Imberger, J.; Garibaldi, L. Summer spatial variations in phytoplankton composition and biomass in surface waters of a warm-temperate, deep, oligo-holomictic lake: Lake Iseo, Italy. *Inland Waters* **2014**, *4*, 303–310. [CrossRef]

42. Miller, A.J. *Subset Selection in Regression*; Chapman & Hall/CRC: Boca Raton, FL, USA, 2002; ISBN 9781584881711.

43. Schützenmeister, A.; Jensen, U.; Piepho, H.P. Checking normality and homoscedasticity in the general linear model using diagnostic plots. *Commun. Stat. Simul. Comput.* **2012**, *41*, 141–154. [CrossRef]

44. R Core Team. *R: A Language and Environment for Statistical Computing*; R Core Team: Vienna, Austria, 2017; ISBN 3_900051_00_3.

45. Wickham, H. *ggplot2: Elegant Graphics for Data Analysis*; Springer-Verlag: New York, NY, USA, 2009; Volume 35, ISBN 9780387981406.

46. Wei, T.; Simko, V. R Package "Corrplot": Visualization of a Correlation Matrix (Version 0.84). Available online: https://github.com/taiyun/corrplot (accessed on 14 November 2017).

47. Kassambara, A.; Mundt, F. Factoextra: Extract and Visualize the Results of Multivariate Data Analyses (Version 1.0.5). Available online: https://CRAN.R-project.org/package=factoextra (accessed on 22 August 2017).

48. Camargo, J.A.; Alonso, Á. Ecological and toxicological effects of inorganic nitrogen pollution in aquatic ecosystems: A global assessment. *Environ. Int.* **2006**, *32*, 831–849. [CrossRef] [PubMed]

49. Rabalais, N.N. Nitrogen in Aquatic Ecosystems. *Ambio* **2002**, *31*, 102–112. [CrossRef] [PubMed]

50. U.S. Environmental Protection Agency. *Aquatic Life Ambient Water Quality Criteria for Ammonia—Freshwater*; U.S. Environmental Protection Agency: Washington, DC, USA, 2013; Volume 78, pp. 67–79.

51. Netten, J.J.C.; van der Heide, T.; Smolders, A.J.P. Interactive effects of pH, temperature and light during ammonia toxicity events in *Elodea canadensis*. *Chem. Ecol.* **2013**, *29*, 448–458. [CrossRef]

52. Loken, L.C.; Small, G.E.; Finlay, J.C.; Sterner, R.W.; Stanley, E.H. Nitrogen cycling in a freshwater estuary. *Biogeochemistry* **2016**, *127*, 199–216. [CrossRef]

53. Gobler, C.J.; Burkholder, J.M.; Davis, T.W.; Harke, M.J.; Johengen, T.; Stow, C.A.; Van De Waal, D.B. The dual role of nitrogen supply in controlling the growth and toxicity of cyanobacterial blooms. *Harmful Algae* **2016**, *54*, 87–97. [CrossRef] [PubMed]

54. Nava, V.; Patelli, M.; Soler, V.; Leoni, B. Interspecific relationship and ecological requirements of two potentially harmful cyanobacteria in a Deep South-Alpine Lake (L. Iseo, I). *Water (Switzerland)* **2017**, *9*, 993. [CrossRef]

55. Glibert, P.M.; Wilkerson, F.P.; Dugdale, R.C.; Raven, J.A.; Dupont, C.L.; Leavitt, P.R.; Parker, A.E.; Burkholder, J.M.; Kana, T.M. Pluses and minuses of ammonium and nitrate uptake and assimilation by phytoplankton and implications for productivity and community composition, with emphasis on nitrogen-enriched conditions. *Limnol. Oceanogr.* **2016**, *61*, 165–197. [CrossRef]

56. Nimptsch, J.; Pflugmacher, S. Ammonia triggers the promotion of oxidative stress in the aquatic macrophyte *Myriophyllum mattogrossense*. *Chemosphere* **2007**, *66*, 708–714. [CrossRef] [PubMed]

57. Gerletti, M.; Provini, A. Effect of nitrification in Lake Orta. In *Ninth International Conference on Water Pollution Research, Proceedings of the 9th International Conference, Stockholm, Sweden, 12–16 June 1978*; Elsevier: Amsterdam, The Netherlands, 1979; pp. 839–851, ISBN 9780080229393.

58. Kim, J.H.; Guo, X.; Park, H.S. Comparison study of the effects of temperature and free ammonia concentration on nitrification and nitrite accumulation. *Process Biochem.* **2008**, *43*, 154–160. [CrossRef]

59. Anthonisen, A.; Loehr, R.; Prakasam, T.; Srinath, E. Inhibition of Nitrification by Ammonia and Nitrous Acid. *J. Water Pollut. Control Fed.* **1976**, *48*, 835–852. [CrossRef] [PubMed]

60. Hira, D.; Aiko, N.; Yabuki, Y.; Fujii, T. Impact of aerobic acclimation on the nitrification performance and microbial community of landfill leachate sludge. *J. Environ. Manag.* **2018**, *209*, 188–194. [CrossRef] [PubMed]

61. Elser, J.J.; Bracken, M.E.S.; Cleland, E.E.; Gruner, D.S.; Harpole, W.S.; Hillebrand, H.; Ngai, J.T.; Seabloom, E.W.; Shurin, J.B.; Smith, J.E. Global analysis of nitrogen and phosphorus limitation of primary producers in freshwater, marine and terrestrial ecosystems. *Ecol. Lett.* **2007**, *10*, 1135–1142. [CrossRef] [PubMed]

62. Yang, Y.; Zhang, J.; Zhao, Q.; Zhou, Q.; Li, N.; Wang, Y.; Xie, S.; Liu, Y. Sediment Ammonia-Oxidizing Microorganisms in Two Plateau Freshwater Lakes at Different Trophic States. *Microb. Ecol.* **2016**, *71*, 257–265. [CrossRef] [PubMed]

63. Ruiz, G.; Jeison, D.; Chamy, R. Nitrification with high nitrite accumulation for the treatment of wastewater with high ammonia concentration. *Water Res.* **2003**, *37*, 1371–1377. [CrossRef]

64. Biesterfeld, S.; Farmer, G.; Russell, P.; Figueroa, L. Effect of alkalinity type and concentration on nitrifying biofilm activity. *Water Environ. Res.* **2003**, *75*, 196–204. [CrossRef] [PubMed]

65. Shanahan, J.W.; Semmens, M.J. Alkalinity and pH effects on nitrification in a membrane aerated bioreactor: An experimental and model analysis. *Water Res.* **2015**, *74*, 10–22. [CrossRef] [PubMed]

66. Li, B.; Irvin, S. The comparison of alkalinity and ORP as indicators for nitrification and denitrification in a sequencing batch reactor (SBR). *Biochem. Eng. J.* **2007**, *34*, 248–255. [CrossRef]

67. Lawson, T.B. *Fundamentals of Aquacultural Engineering*; Springer: Boston, MA, USA, 1995; ISBN 978-1-4612-7578-7.

68. Isnansetyo, A.; Getsu, S.; Seguchi, M.; Koriyama, M. Independent Effects of Temperature, Salinity, Ammonium Concentration and pH on Nitrification Rate of the Ariake Seawater Above Mud Sediment. *HAYATI J. Biosci.* **2014**, *21*, 21–30. [CrossRef]

Article

Effects of Water Level Increase on Phytoplankton Assemblages in a Drinking Water Reservoir

Yangdong Pan [1,2], Shijun Guo [1,3], Yuying Li [1,3,*], Wei Yin [1,4], Pengcheng Qi [1,3], Jianwei Shi [1,5], Lanqun Hu [1,5], Bing Li [1,3], Shengge Bi [1,3] and Jingya Zhu [1,3]

[1] Collaborative Innovation Center of Water Security for Water Source Region of Middle Route Project of South-North Water Diversion of Henan Province, Nanyang Normal University, Nanyang 473061, Henan, China; pany@pdx.edu (Y.P.); 152256558851@163.com (S.G.); 2000yinwei@163.com (W.Y.); qipengchengsd@163.com (P.Q.); nyhbjsjw@126.com (J.S.); hlq65@sina.com (L.H.); benfei111@163.com (B.L.); 15890033526@163.com (S.B.); zhujy15@126.com (J.Z.)
[2] Department of Environmental Science and Management, Portland State University, Portland, OR 97207, USA
[3] Key Laboratory of Ecological Security for Water Source Region of Middle Route Project of South-North Water Diversion of Henan Province, College of Agricultural Engineering, Nanyang Normal University, Nanyang 473061, Henan, China
[4] Changjiang Water Resources Protection Institute, Changjiang Water Resources Commission, Wuhan 430051, Hubei, China
[5] Emergency Centre for Environmental Monitoring of the Canal Head of Middle Route Project of South-North Water Division, Xichuan 474475, Henan, China
* Correspondence: lyying200508@163.com; Tel.: +86-377-6352-5027

Received: 18 January 2018; Accepted: 27 February 2018; Published: 2 March 2018

Abstract: Excessive water level fluctuation may affect physico-chemical characteristics, and consequently ecosystem function, in lakes and reservoirs. In this study, we assessed the changes of phytoplankton assemblages in response to water level increase in the Danjiangkou Reservoir, one of the largest drinking water reservoirs in Asia. The water level increased from a low of 137 m to 161 m in 2014 as a part of the South–North Water Diversion Project. Phytoplankton assemblages were sampled four times per year before, during and after the water level increase, at 10 sites. Environmental variables such as total nitrogen as well as phytoplankton biomass decreased after the water level increase. Non-metric multi-dimensional scaling analysis indicated that before the water level increase, phytoplankton assemblages showed distinct seasonal variation with diatom dominance in both early and late seasons while such seasonal variation was much less evident after the water level increase. Month and year (before and after) explained 13% and 6% of variance in phytoplankton assemblages (PERMANOVA, $p < 0.001$) respectively, and phytoplankton assemblages were significantly different before and after the water level increase. Both chlorophytes and cyanobacteria became more abundant in 2015. Phytoplankton compositional change may largely reflect the environmental changes, such as hydrodynamics mediated by the water level increase.

Keywords: Danjiangkou Reservoir; South–North Water Diversion Project; non-metric multi-dimensional scaling (NMDS); PERMANOVA

---

1. Introduction

Ecological impacts of excessive water level fluctuation (WLF) in both lakes and reservoirs have received increasing attention from both ecologists and water resource managers [1–5]. The water level does fluctuate naturally, but an increase in global climate variability (e.g., precipitation) is expected to result in more extreme hydrological events, such as extended droughts and flooding. Consequently, both intensity and frequency of excessive WLF in lakes and reservoirs will increase, which will adversely affect water supply, food security, and aquatic ecosystems [6–8]. Excessive WLF may affect both physical and chemical characteristics of lakes and reservoirs including quantity and quality of littoral zone habitat, light, thermal stratification/mixing, and internal nutrients and consequently affect freshwater ecosystem structure and function [8–11]. For instance, water level fluctuation caused significant reduction of benthic algal productivity and macroinvertebrate diversity in littoral zones [8]. The littoral zones, a critical ecotone for biodiversity, habitats, primary productivity, and material exchanges between aquatic and terrestrial ecosystems [10], have been the focus of many ecological studies on the impacts of WLF, particularly in shallow lakes and wetlands, two of the most vulnerable ecosystems to WLF [9]. The studies on the ecological impacts of WLF on deep drinking water reservoirs are still relatively limited [8].

The intensity and frequency of WLF in deep reservoirs in response to anthropogenic demands—such as drinking water, hydropower generation, or irrigation—may be much higher than that in natural lakes, particularly under the projected extreme climate change scenarios [8]. For instance, the maximum amplitude of WLF in Lake Shasta, a reservoir in California (USA), reached as high as 47 m between 1991 and 1993 compared to less than 3 m in some natural lakes [2]. In deep lakes and reservoirs, excessive WLF can also potentially disrupt thermal structure, alter mixing regimes and internal nutrient cycles, and consequently increase the likelihood for algal blooms [1,5,12,13]. For instance, due to an extended drought, the water level in Lake Burragorang—a drinking water reservoir in Australia—dropped 25 m. Vilhena et al. used both hydrodynamic and ecological models to demonstrate that large river inflow could disrupt thermal stratification when the lake water level was extremely low, release dissolved nutrients from the hypolimnion, and result in a subsequent *Microcystis* bloom. It is expected that climate change will likely aggravate the negative impacts of WLF on drinking water reservoirs, particularly in the regions with high population growth, urbanization, and rapid economic development [8]. Better understanding of the impacts of excessive WLF on drinking water reservoirs, particularly on water quality, will help water resource managers design optimal operation and develop mitigation plans for climate change.

Existing water infrastructure are often not designed to cope with the impacts of climate change [1,7]. China's South–North Water Diversion Project is a part of its national strategy to adapt and mitigate climate change and meet a rapidly increasing demand for drink water and irrigation. This multi-billion dollar project aims to divert abundant water resource from the Yangtze River and its tributaries in the south to the water-stressed north China plain. The Danjiangkou Reservoir, one of the largest drinking water reservoirs in Asia, is the source water for the middle route of the South–North Water Diversion Project [14]. To prepare for the water diversion, the water level at the reservoir was increased after the original dam height was raised to increase its water storage capacity. This large-scale water-level manipulation in a reservoir may provide a valuable natural experiment on the ecological effects of WLF and future best management practices. In this study, we tested if the water level increase had significant effects on phytoplankton assemblages in the reservoir before and after the water level increase. Most of the studies on WLF focus on the effects of decreasing water levels on aquatic ecosystems. The studies on the ecological impacts of excessive WLF on drinking reservoirs are often hampered by limited long-term monitoring data of physio-chemical and particular biological parameters. Under the projected extreme climate change scenarios, it is expected that drinking water reservoirs including Danjiangkou may experience higher intensity and frequency of water level fluctuation. Both design and operation of water infrastructure such as drinking reservoirs traditionally focus more on water quantity. A better operation and management practice should include water quality as one of the management priorities, particularly because several excessive WLF events resulted in harmful algal blooms [1,5,12,13]. Characterizing phytoplankton assemblages with relation to water level changes will help develop a long-term ecological monitoring system for drinking water quality in the future. We expected that the increase of the water level in the reservoir would affect both hydrodynamics and nutrients associated with newly flooded lands and consequently both spatial and temporal variation of phytoplankton assemblages in the reservoir.

2. Materials and Methods

2.1. Study Area

The Danjiangkou Reservoir, located in central China (32°36′–33°48′ N; 110°59′–111°49′ E), is one of the largest reservoirs in Asia (Figure 1). The dam was initially constructed on the upper reach of the Han River, one of the largest tributaries of the Yangtze River, in 1973. As one of China's three trans-basin water transfer projects [14], the reservoir's water storage capacity was further expanded by increasing the dam height from 162.0 m to 176.6 m above sea level in 2013 (Table A1). The reservoir surface area was increased from 745 km^2 at the normal water level of 157 m to 1050 km^2 at the normal water level of 170 m with a maximum water depth of approximately 80 m. The Han and Dan rivers, two major tributaries, flow into the reservoir and form two major arms, i.e., the Han Reservoir (DRH) and the Dan Reservoir (DRD), separated by the provincial border line between Hubei and Henan provinces (Figure 1). The reservoir is approximately isothermal from November to March and thermal stratification typically starts in April [15].

The climate in the region is characterized as a subtropical monsoon with hot and humid summers and mild winters. The mean annual temperature is 15.7 °C, with a monthly average of 27.3 °C in July and 4.2 °C in January. Approximately 80% the annual precipitation of 749.3 mm occurs between May and September (wet season) [16]. With >200 tributaries flowing into the reservoir, the watershed includes an area of approximately 95,200 km^2. The elevation in the watershed varies between 201 and 3500 m a.s.l. The watershed is dominated by forests (>75%) with major vegetation types varied from deciduous broadleaves, conifers, to alpine shrub meadows along an elevational gradient [17]. Approximately 15% of the watershed, primarily in the area with a gentle slope below an elevation of 1000 m, is agricultural land with corn and wheat as the two major crops. The top 30 cm of soil consists of 48% clay, 41% silt, and 11% sand [18].

2.2. Field Sampling and Sampling Time

A total of 10 sites were selected for assessing spatial variability of phytoplankton assemblages in the reservoir (Figure 1b) and each site was sampled in January, May, July, and October during 2014–2016 for assessing both seasonal and year-to-year variability. The three years corresponding to the three phases of the water level changes. The number of sites, skewed more toward the Dan River arm (DRD) (n = 7) than the Han River arm (n = 3), reflect the spatial variability within each arm of the reservoir. The Han River arm, confined by narrow rocky canyons, is much more river-like with relatively fast flow while the Dan River arm includes both upper riverine section and a much more lentic pelagic zone below the confluence of the Dan and Guan rivers (Figure 1b). At each site, a sample of 5 L of water for phytoplankton assemblage was collected at 50 cm below the water surface using a Van Dorn sampler and was preserved immediately with 1% Lugol's solution in January, May, July, and October during the period of 2014–2016. Water temperature (T), pH, conductivity (CD), and dissolved oxygen (DO) were measured in situ using a YSI multi-probe (YSI 6920, USA). Transparency was measured using a Secchi disk (SD).

2.3. Laboratory Analysis

Each phytoplankton sample was concentrated to 30 mL using a glass sedimentation utensil for 48 h. Phytoplankton was counted in a Fuchs–Rosental counting chamber of 0.1 mL under a microscope (Nikon E200, Kanagawa ken, Japan) at 400× magnification. A minimum of 500 algal units were identified to the lowest taxonomic level possible and enumerated for characterizing overall algal assemblages [19–21]. Biovolume of each taxon was calculated based on measured morphometric characteristics (diameter, length, and width). Conversion to biomass was based on 1 mm^3 of algal volume equals 1 mg of fresh weight biomass. For diatom species identification, samples were processed with acid reagents [19].

Total phosphorus (TP) and total nitrogen (TN) were analyzed based on the standard methods of Chinese Ministry of Environmental Protection [22,23]. TP was analyzed using ammonium molybdate spectrophotometric method while TN was analyzed using alkaline potassium persulfate digestion UV spectrophotometric method. Chemical oxygen demand (COD) was analyzed with the potassium dichromate method.

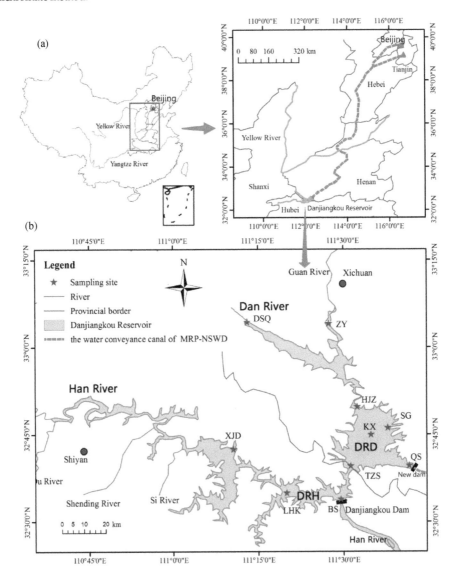

Figure 1. Map of China showing (**a**) water diversion route from the reservoir to the destination cities (Beijing and Tianjin); and (**b**) sampling sites in the Danjiangkou Reservoir. The map was generated using ArcGIS 10.0 (ESRI, Redlands, CA, USA: http://www.esri.com/software/arcgis).

2.4. Data Analysis

To summarize both spatial and temporal variation of phytoplankton assemblages in the reservoir, we performed non-metric multidimensional scaling (NMDS) with Bray–Curtis dissimilarity index [24]. Algal data as relative abundance of total algal biomass (n = 118, taxa = 193) were square-root transformed to dampen the impacts of dominant taxa on the ordination analysis. The NMDS was run 20 times each with a random starting configuration. The final NMDS dimension was selected based on the lowest stress value among the best solutions. The effects of the reservoir arms (DRH vs. DRD), year (before vs. after water-level increase), and month on phytoplankton assemblages were assessed using permutational multivariate analysis of variance (PERMANOVA, 'adonis' function in 'Vegan' R package [25]) with repeated measures (month), a non-parametric multivariate statistical test [26]. PERMANOVAs were then conducted for each sampling month with Bonferroni's correction. Prior to the PERMANOVA tests, the homogeneity of multivariate dispersions among the groups was assessed with Bray–Curtis dissimilarity measure using the 'betadisper' function in 'vegan' of R package [25]. The test for the homogeneity of multivariate dispersions (Anderson 2006) is a multivariate analogue to Levene's test [27]. In addition, the phytoplankton assemblage distribution patterns captured by the NMDS were also related with measured environmental variables using 'envfit' function in 'Vegan' R package [25]. This function fits explanatory variables in the ordination space defined by the species data (i.e., NMDS plot). The importance of each vector was assessed using a squared correlation coefficient (R^2). Finally, to characterize what taxa were largely responsible for the assemblage-level changes before, during vs. after water-level increase, the indicator taxa for each year were identified following the methods by Dufrene & Legendre (1997). An indicator value of a species for a year was the product of the relative abundance and the relative frequency of the taxa in the year. The statistical significance of the indicator value was then tested using a permutation test (999 permutations). All analyses were performed using R [28] and an R packages of 'vegan' [25] and 'labdsv'. The datasets generated and/or analyzed during the current study are available from the corresponding author on reasonable request.

3. Results

3.1. Changes of Water Level and Water Quality

Mean annual water level changed from 144 m (2012–2013) to 154 m (2015–2016) in the Danjiangkou Reservoir (Figure 2, Table A1). Between April and December 2014, the water level increased from 137 m to 161 m, an increase of 24 m (mean = 10 cm/day). From April to October, the water level increased as a part of seasonal variation while from October to December, the water level continued to increase as a part of the water transfer project, which started to release water to the distributional canals on 12 December 2014. The water levels showed pronounced wet and dry seasonal cycles in 2012 and 2013. The annual variations were 17 and 13 m, respectively. After the water transfer project started, annual variation decreased substantially (7 m in 2015 and 6 m in 2016) with no distinct wet and dry seasonal cycles.

Water in the reservoir in general had high conductivity, low to moderate nutrients, and high water transparency (Figure A1). Water temperature showed similar seasonal variation before, during and after the water level increase. Water quality variables did not show consistent changes before and after the water level increase (Figure A1). Median specific conductivity increased from 336.5 μS/cm in 2014 to 368.5 μS/cm and 342.0 μS/cm in 2015 and 2016, respectively while both median TN and TP concentrations decreased (TN: 2.21 mg/L in 2014, 1.90 mg/L in 2015 and 1.21 mg/L in 2016; TP: 0.05 mg/L in 2014, 0.03 mg/L in 2015, and 0.02 mg/L in 2016). Monthly spatial variation of TN was higher in both 2015 and 2016 than in 2014. Median water transparency remained high (>3.3 m) before and after the water level increase. Median pH were 7.38, 7.43, and 7.33 while median COD was 13.60, 12.95, and 13.20 mg/L before, during, and after the water level increase.

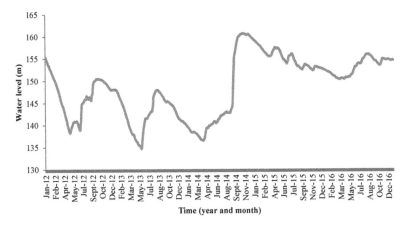

Figure 2. Daily water level changes in the Danjiangkou Reservoir (near the Danjiangkou Reservoir Dam) between 2012 and 2016.

3.2. Changes of Phytoplankton Assemblages in Relation to Environmental Variables

Phytoplankton assemblages changed seasonally before water level increase and year-to-year before and after water level changes (Figure 3). The results of PERMANOVA with repeated measures show that month and year (before, during, and after) as factors explained 13% and 6% of variance in phytoplankton assemblages ($p < 0.001$ for both factors), respectively, while spatial variation between the two reservoir arms only explained 1% of variance in phytoplankton assemblages. Median phytoplankton biomass in 2014 was 7.4 mg/L and then decreased to 4.9 mg/L and 2.3 mg/L in 2015 and 2016, respectively (Figure A2). PERMANOVA of phytoplankton assemblages for each month indicated that phytoplankton assemblages were significantly different before, during, and after the water level increase for each sampling month ($p < 0.001$) (Table A2). Before the water level increase, diatoms in both dominance and species composition shifted from a cold and dry season (January and May) to a warm and wet season (July and October) along the first NMDS axis (Figure 3 and Figure A3). In 2014, median diatom levels were highest at 70% of phytoplankton biomass in January and dominated by *Cyclotella* sp. (30.74%) and *Aulacoseira granulata* (Ehr.) Simonsen (30.66%) (Figure A3a, Table A3). Dominance of diatoms continued to decline in May and reached its lowest of 20% in July as median chlorophytes (e.g., *Eudorina* sp.) reached 30% (except for two sites) (Figure 3b). Diatoms became dominant again in October but mainly by *Cymbella* sp. (14.2%), *Synedra* sp. (14.0%), and *Cyclotella* sp. (13.0%) (Figure A3a, Table A3).

Most noticeable changes after the water level increase were that the seasonal differences in phytoplankton assemblages between the dry/cold and warm/wet seasons in 2015 and 2016 became much less evident than in 2014 (Figure 3a). Median chlorophyte relative abundance increased from 8.5% in 2014 to 19.7% and 21.1% in 2015 and 2016, an increase of 57% and 60%, respectively. Chlorophytes were abundant in the early season (May) in 2015, particularly at the sites XJD, BS, and LHK where relative abundance was >80% (Figure A3b). Overall cyanobacteria was relatively low in abundance in 2014 with the highest of 17% at site KX but increased in median abundance to 40% in October 2015 (Figure A3c). At sites TZS and BS, relative abundance was 78% and 72%, respectively, compared to 3.3% and 7.0% in 2014. A similar pattern was also observed in 2016 but overall relative abundance in October was substantially lower (median 10%) except for two sites with 40% and 70%, respectively. *Lyngbya* sp., a filamentous benthic cyanobacteria with no heterocysts, was the dominant taxon (Table A3). The indicator taxa of phytoplankton assemblages before, during, and after water level increase were presented in Table A4.

NMDS axis II may primarily reflect the changes in phytoplankton assemblages after the water level increase, particularly in October. The second axis separates phytoplankton assemblages dominated by cyanobacteria (Figure 3c) from chlorophytes (Figure 3d). The analysis of vector-fitting of environmental variables to the NMDS space indicated that several environmental variables significantly correlated with the ordination space defined by phytoplankton species composition. Both seasonal variation and before-and-after water level changes in phytoplankton assemblages were primarily correlated with water temperature ($R^2 = 0.44$, $p < 0.001$). Correlation with several other variables including COD, pH, and TP were statistically significant ($p < 0.05$) but the correlation was relatively weak ($R^2 < 0.1$) (Figure 3a).

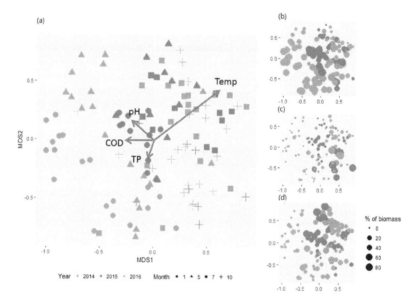

Figure 3. Site plot of non-metric multidimensional scaling (NMDS) analysis showing (**a**) temporal and spatial variation of phytoplankton assemblages between 2014 and 2016 in the Danjiangkou Reservoir with relation to major environmental variables; (**b–d**) the same plot superimposed with % of diatoms, cyanobacteria, and chlorophytes in each sample (bubble size is proportional to the relative abundance of total phytoplankton biomass).

4. Discussion

Water level increase in the Danjiangkou Reservoir did not result in algal blooms. It was expected that, as water level increased as a result of precipitation, runoff and river inflow, turbidity, and nutrients from both river inflow/runoff and newly inundated land would increase to a level which may lead to an increase in algal and macrophyte production, and in some cases, algal blooms [2,9]. In contrast, median phytoplankton biomass as well as nutrients (TN and TP) in the reservoir decreased during and after the water level increase (2015 and 2016) while water transparency remained high before, during, and after the water level increase. As several researchers cautioned, physio-chemical and ecological processes associated with WLF can be complex [2,29]. The amplitude, increasing rate, and timing of WLF as well as the characteristics of lakes/reservoirs (e.g., water depth) and watersheds (e.g., topography and land-use) can all affect how a lake/reservoir may respond to water level changes. For instance, Lake Burragorang in Australia experienced three major extended droughts and WLFs between 1988 and 2008, but WLFs only resulted a major algal bloom once [13]. Vilhena et al. used both hydrodynamic and ecological models to demonstrate that a combination of the low water level in the lake and the high river inflow could alter both mixing regime and internal nutrient cycling, which could lead to a *Microcystis* bloom. Refilling a completely empty Brucher Reservoir in central-western Germany was followed by nutrient enrichment as a result of decomposition of submerged terrestrial vegetation [30]. Due to complete drawdown, zooplankton with no predation pressure grazed down increased algal biomass, which prevented algal blooms during the initial refilling process. In four Sardinian reservoirs (Italy), increasing levels of eutrophication and the abundance of cyanobacteria occurred due to trophic, hydrological, and seasonal patterns, especially in the southern Mediterranean, and the most important species had significant correlations with nutrients and microcystins [31]. Lack of algal blooms in response to water level increase in the Danjiangkou Reservoir may be attributed to several factors. A 10 m increase of the water level inundated approximately 176.75 km^2 of new land around the reservoir including a large portion of agricultural land. The nutrients leached from the submerged land may be short-lived (e.g., days or weeks) [32,33] and not high enough to alter the nutrient concentrations in the water column. Gordon et al. experimentally compared the effects of rewetting dry soils from grassland with low and high productivity and showed nutrient leaching largely depended on soil productivity [34]. Our data showed that both nutrients (TN and TP) actually decreased after the water level increase, which may be due to the combination of relatively poor soil quality with approximately 60% of top soil as clay and sand [18] and the dilution factor of the large reservoir since the overall inflow volume was still relatively small compared to the volume of the reservoir. The water level was increased during the period when the reservoir was approximately isothermal and incoming nutrients may be mixed through the entire water column. Chen et al. modelled the thermal regimes in the reservoir and their model indicated that the thermal stratification usually occurred between May and October [15]. Finally, the water level increased during the period when algal biomass may be approaching the so-called "winter minimum" due to reduced light energy [35].

Phytoplankton assemblages did change after the water level increase in the Danjiangkou Reservoir. Before the water level increase, phytoplankton assemblages showed distinct seasonal variation with diatom dominance in both early and late seasons, commonly observed in a deep dimictic waterbody (Figure 3) [36]. Unlike the early season, particularly in January when the assemblages were primarily dominated by two euplanktonic diatom taxa (*Cyclotella* sp. and *A. granulate*), diatoms were still dominant but with a higher proportion of benthic taxa in the late season, particularly in October (Table A2). A higher proportion of benthic diatoms may be from high river inflow during the wet season. It is interesting that NMDS shows that the seasonal variation in phytoplankton assemblages was much less evident after the water level increase (Figure 3). In addition, chlorophytes were abundant in early season (May) in 2015 while cyanobacteria became abundant in October 2015. Compositional change of phytoplankton assemblage may largely reflect changes in environmental conditions, such as newly available littoral zones, hydrodynamics, and thermal regimes mediated by the water level increase. Water level increase at a slow and steady pace may create a "moving littoral zone" where active aquatic-terrestrial interaction may take place [29]. With median water transparency remaining at >3 m, a "moving littoral zone" may provide newly available habitats, particularly for benthic algae which can colonize a wide range of substrates [37]. It is interesting that increased cyanobacteria in October 2015 was mainly attributed to *Lyngbya* sp., a benthic filamentous taxon. Replacement of benthic diatoms in October by cyanobacteria after the water level increase may be attributed to opportunistic nature of *Lyngbya* sp. with the ability to quickly colonize newly available benthic habitats and altered thermal regimes. *Lyngbya* often start as a benthic mat in marginal areas of lakes/reservoirs or large rivers and then develop into floating mats [38,39]. The taxa has a high temperature optimum, low light requirement, and high tolerance for a wide range of nutrient conditions and quickly colonized and established large populations immediately after aquatic macrophyte beds were removed by herbicides in southeastern US lakes [38], or new habitats became available in the Great Lakes (USA) [40] and then became a nuisance as floating mats. The taxa grow best at depth between 1.5 and 3.5 m in mixed substrates of sand and fragmented shells and grew in water with high turbidity and DOC associated with agricultural land in the St. Lawrence River and Maumee River (USA) [39]. Chen et al. used a hydrodynamic model to assess the effects of water level increase on the thermal regimes in the Danjiangkou Reservoir [15]. Their model predicted that the water level increase may increase water temperature in the winter and decrease it in the summer. In addition, the water level increase may also alter vertical thermal stratification. Lack of 'normal' water level fluctuation and associated altered seasonal temperature variation in the reservoir may also contribute to diminished seasonal variation in phytoplankton assemblages.

Climate change poses additional challenges on the management of water infrastructure, such as drinking water reservoirs. Adaptive water management practices and mitigation plans will likely reduce the impact of climate change on water resources as well as adaptation costs [8]. It is critical for reservoir managers to understand how WLF may affect reservoir ecosystems so that a mitigation plan can be developed [5]. It was unexpected that *Lyngbya* sp., a mat-forming benthic cyanobacterial taxa, became so abundant during water level increase. Some species in this genus such as *L. wollei* have been reported to produce a potent, acutely lethal neurotoxin [41]. Unfortunately, in this study, we were unable to positively identify *Lyngbya* sp. to the species level. Further studies on its taxonomic identification including laboratory culture and molecular work, reoccurrence, and persistence are needed to assure drinking water safety. Our results also indicated that pronounced seasonal succession of phytoplankton assemblages was diminished after water level increase in 2016, an indication of changed thermal dynamics in the reservoir. Its negative impacts on several fish species downstream have been assessed and several management plans have been developed to mitigate the negative effects of water level increase on water temperature and downstream biota [15]. It is not clear if abundant *Lyngbya* sp. was associated with changing thermal regimes. Effects of water level increase on ecosystem structure and function in the reservoir are much more complicated than the physico-chemical variables and thus the observed results in this study with limited study duration should be interpreted with caution. It is difficult to discern how much biotic variation is due to natural or human induced WLF since it is almost impossible to have a true natural reference for man-made impoundments [29]. It is equally difficult to assess the long-term effects of WLF on ecosystem structure and function. Reservoir managers should incorporate a long-term biological monitoring plan as an integral part of the drinking water reservoir management plan. Such a long-term biological monitoring plan should also be enhanced by emerging new DNA sequencing technology to assure both accuracy and consistence of taxonomic identification [42].

5. Conclusions

In summary, phytoplankton compositional change may largely reflect the environmental changes such as hydrodynamics mediated by the water level increase. In the drinking water source area, long-term monitoring plans should include biological assemblages to better reflect the impacts of environmental conditions on ecological processes in the waters.

Acknowledgments: The research was supported by the Key Research and Development Program of China (2016YFC0402204 and 2016YFC0402207), the Major Science and Technology Project for Watershed Pollution Control and Management (2017ZX07108), the National Science Foundation of China (41601332 and U1704124), the scientific project of Henan Province (2016151, 17454 and 182102310223), and the key scientific project of Education Department of Henan Province (16A210012). We are grateful to the editor, three anonymous reviewers, and Eugene Foster for their helpful comments and suggestions.

Author Contributions: P.Y.D. and L.Y.Y. conceived the study and wrote the main manuscript text. P.Y.D., L.Y.Y., and Z.J.Y. analyzed the data. Figure 1 was prepared by Q.P.C. and others by P.Y.D. All authors contributed to interpreting the results and editing the manuscript, and gave final approval for publication.

Conflicts of Interest: The authors declare no conflict of interest.

Appendix A

Table A1. Comparison of main characteristics of the Danjiangkou Reservoir before and after the dam heightening.

Fundamental Parameters	Stage	
	Before Dam Heightening	After Dam Heightening
Dam height (m)	162	176.6
Normal water level (m)	157	170
Total capacity (billion m^3)	174.5	290.5
Dead water level (m)	140	150
Adjustable storage capacity (billion m^3)	98.0~102.2	136.6~190.5
Reservoir area (km^2)	745	1050
Limit drop depth (m)	18	25
Backwater length (km) DRH	177	193.6
Backwater length (km) DRD	80	93.5
Bank shoreline length (km)	4600	7000
Capacity ratio	0.45	0.75

Table A2. Results of permutational multivariate analysis of variance (PERMANOVA) of phytoplankton assemblages for each sampling month. The bold *p*-values are statistically significant after Bonferroni correction.

Variance Index	Df	SS	MS	F value	R^2	*p* Value
			January			
Reservoir arm	1	0.4923	0.4923	2.1661	0.06	0.022
Time	1	2.209	2.2090	9.7196	0.25	**0.001**
Residuals	27	6.1364	0.2273		0.69	
			May			
Reservoir arm		0.5764	0.5764	2.1475	0.06	0.03
Time		2.5646	2.5646	9.5547	0.25	**0.001**
Residuals		7.2471	0.2684		0.7	
			July			
Reservoir arm	1	0.3205	0.3205	1.0795	0.03	0.37
Time	1	1.9596	1.9596	6.5992	0.2	**0.001**
Residuals	26	7.7207	0.297		0.77	
			October			
Reservoir arm	1	0.2965	0.2965	0.9312	0.03	0.499
Time	1	1.3193	1.3193	4.1428	0.13	**0.001**
Residuals	26	8.2802	0.3185		0.84	

Table A3. Seasonal and year-to-year variation of mean relative abundance of the dominant taxa (% of total biomass) in the Danjiangkou Reservoir between 2014 and 2016.

Taxa	Division	2014				2015				2016			
		1	5	7	10	1	5	7	10	1	5	7	10
Aulacoseira granulata (Ehr.) Simonsen	Bacillariophyta	30.66	5.86	0.19	0.92	7.42	0.00	0.00	0.27	11.53	17.70	2.24	0.56
Cyclotella sp.	Bacillariophyta	30.74	10.83	11.59	12.95	25.57	5.07	7.46	2.72	1.90	4.20	1.02	9.72
Cymbella sp.	Bacillariophyta	0.00	0.00	0.00	14.20	0.32	0.00	6.47	0.00	0.00	0.00	0.00	0.00
Melosira varians Agardh	Bacillariophyta	0.39	0.18	0.16	0.74	1.41	0.50	0.04	0.00	6.82	11.14	1.73	0.71
Navicula sp.	Bacillariophyta	0.13	0.92	4.31	0.28	1.59	8.58	0.00	0.00	5.10	10.00	4.36	2.52
Synedra sp.	Bacillariophyta	1.65	0.00	3.96	14.00	3.01	0.25	1.01	12.35	9.78	1.46	4.86	4.43
Chlamydomonas sp.	Chlorophyta	0.48	0.00	0.00	0.00	1.76	12.67	8.63	0.00	0.45	0.00	0.50	0.00
Eudorina elegans Ehr.	Chlorophyta	0.00	1.95	4.00	7.34	6.05	21.24	0.00	5.42	0.00	6.26	3.50	11.63
Eudorina sp.	Chlorophyta	0.00	0.00	14.80	4.40	0.00	5.35	0.45	1.05	0.00	0.00	1.83	0.00
Scenedesmus sp.	Chlorophyta	0.00	0.00	3.59	3.40	1.36	0.55	1.13	0.00	0.00	2.41	3.67	11.24
Lyngbya sp.	Cyanophyta	0.01	0.00	0.00	0.00	0.06	0.01	0.01	26.78	0.83	0.01	0.74	13.55
Trachelomonas sp.	Euglenophyta	2.01	24.89	6.04	0.12	2.67	0.34	0.25	0.24	0.00	0.00	0.36	0.00
Ceratium hirundinella (Müll.) Schr.	Pyrrophyta	0.00	0.00	0.00	1.34	0.00	0.00	9.33	0.00	0.00	3.08	22.90	0.00
Peridinium sp.	Pyrrophyta	6.95	12.36	0.00	0.00	0.00	0.00	0.00	0.00	0.00	0.00	0.00	0.00

Table A4. Indicator taxa of phytoplankton assemblages before, during, and after water level increase in the Danjiangkou Reservoir.

Taxa	Division	Year	Water Level Change	Indicator Value	*p* Value
Trachelomonas sp.	Euglenophyta	2014	Before	0.59	0.001
Aulacoseira granulata var. *angustissima* (Müll.) Simonsen	Bacillariophyta	2014	Before	0.48	0.001
Peridinium sp.	Pyrrophyta	2014	Before	0.27	0.001
Eudorina sp.	Chlorophyta	2014	Before	0.25	0.001
Ceratium sp.	Pyrrophyta	2014	Before	0.12	0.033
Cryptomonas erosa Ehr.	Cryptophyta	2015	During	0.55	0.001
Chlamydomonas sp.	Chlorophyta	2015	During	0.41	0.001
Cyclotella sp.	Bacillariophyta	2015	During	0.36	0.029
Merismopedia elegans Braun.	Cyanophyta	2015	During	0.17	0.041
Fragilaria sp.	Bacillariophyta	2016	After	0.49	0.001
Melosira varians Agardh	Bacillariophyta	2016	After	0.43	0.001
Aulacoseira granulata (Ehr.) Simonsen	Bacillariophyta	2016	After	0.29	0.007
Pediastrum sp.	Chlorophyta	2016	After	0.08	0.023
Navicula sp.	Bacillariophyta	2016	After	0.30	0.046

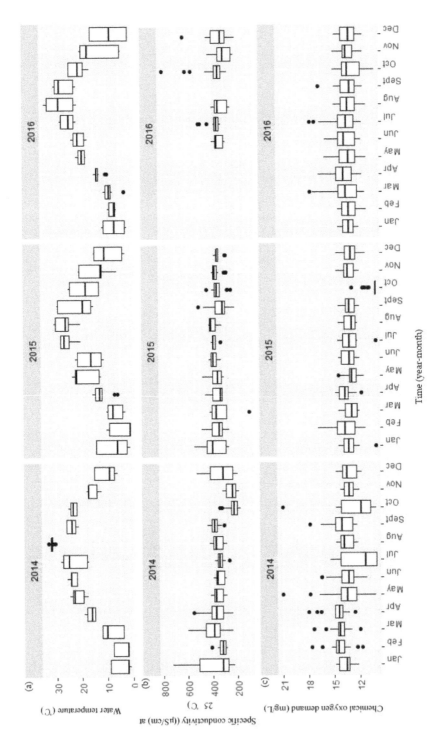

Figure A1. *Cont.*

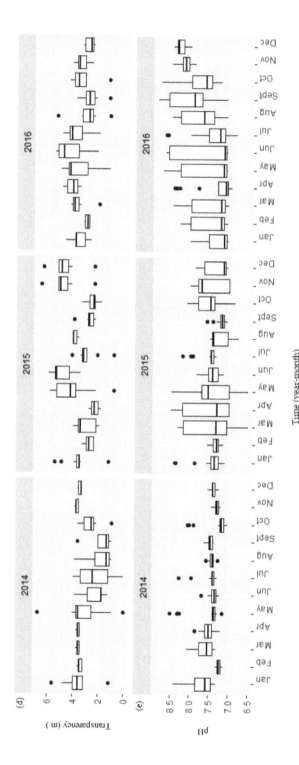

Figure A1. *Cont.*

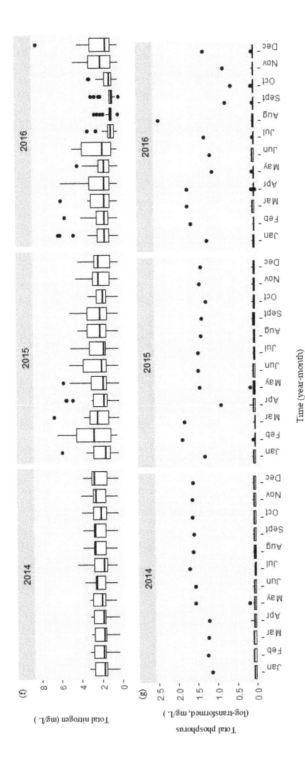

Figure A1. Boxplots showing both temporal and spatial variation of the selected environmental variables (a–g) between 2014 and 2016 in the Danjiangkou Reservoir. (a–g) boxplots were the plots of T, CD, COD, SD, pH, TN, and TP variables, respectively. The vertical line includes a range of variation and the data points outside the range indicated as solid dots are considered as outliers. The box includes the data between first and third quantiles and the median (black bar). The horizontal black bar in Figure A1b indicates missing conductivity data between January and May 2016.

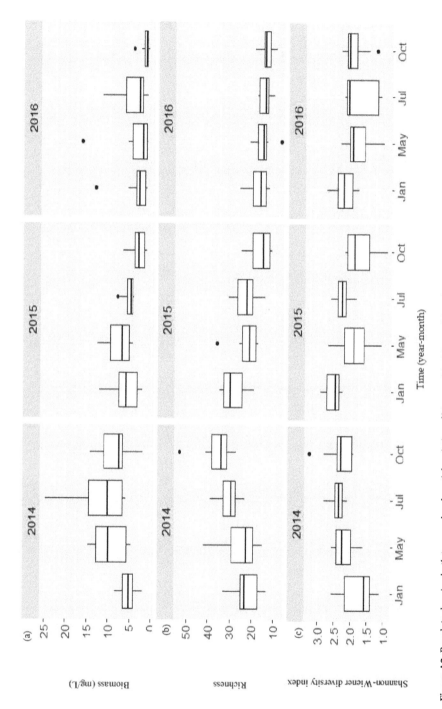

Figure A2. Boxplots showing both temporal and spatial variation of biomass (**a**), richness (**b**), and Shannon–Wiener diversity index (**c**) of phytoplankton assemblages between 2014 and 2016 in the Danjiangkou Reservoir.

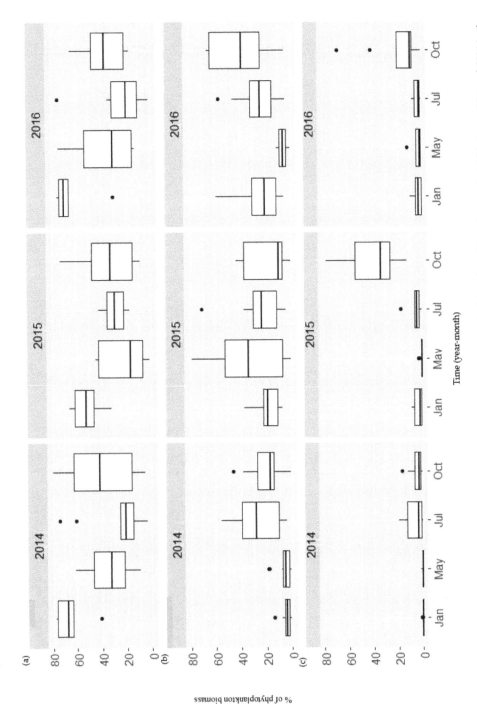

Figure A3. Boxplots showing both temporal and spatial variation of bacillariophyta (**a**), chlorophyta (**b**), and cyanophyt (**c**) between 2014 and 2016 in the Danjiangkou Reservoir.

References

1. Naselli-Flores, L.; Barone, R. Water-level fluctuations in Mediterranean reservoirs: Setting a dewatering threshold as a management tool to improve water quality. *Hydrobiologia* **2005**, *548*, 85–99. [CrossRef]
2. Zohary, T.; Ostrovsky, I. Ecological impacts of excessive water level fluctuations in stratified freshwater lakes. *Inland Waters* **2011**, *1*, 47–59. [CrossRef]
3. Qian, K.M.; Liu, X.; Chen, Y.W. Effects of water level fluctuation on phytoplankton succession in Poyang Lake, China—A five year study. *Ecohydrol. Hydrobiol.* **2016**, *16*, 175–184. [CrossRef]
4. Ji, D.B.; Wells, S.A.; Yang, Z.J.; Liu, D.F.; Huang, Y.L.; Ma, J.; Berger, C.J. Impacts of water level rise on algal bloom prevention in the tributary of Three Gorges Reservoir, China. *Ecol. Eng.* **2017**, *98*, 70–81. [CrossRef]
5. Naselli-Flores, L. Mediterranean climate and eutrophication of reservoirs: Limnological skills to improve management. In *Eutrophication: Causes, Consequences and Control*; Ansari, A., Singh Gill, S., Lanza, G., Rast, W., Eds.; Springer: Dordrecht, The Netherlands, 2010.
6. Bond, N.R.; Lake, P.S.; Arthington, A.H. The impacts of drought on freshwater ecosystems: An Australian perspective. *Hydrobiologia* **2008**, *600*, 3–16. [CrossRef]
7. Bates, B. *Climate Change and Water: IPCC Technical Paper VI*; World Health Organization: Geneva, Switzerland, 2009.
8. Evtimova, V.V.; Donohue, I. Quantifying ecological responses to amplified water level fluctuations in standing waters: An experimental approach. *J. Appl. Ecol.* **2014**, *51*, 1282–1291. [CrossRef]
9. Leira, M.; Cantonati, M. Effects of water-level fluctuations on lakes: An annotated bibliography. *Hydrobiologia* **2008**, *613*, 171–184. [CrossRef]
10. Strayer, D.L.; Findlay, S.E. Ecology of freshwater shore zones. *Aquat. Sci.* **2010**, *72*, 127–163. [CrossRef]
11. Wolin, J.A.; Stone, J.R. Diatoms as indicators of water-level change in freshwater lakes. In *The Diatoms: Applications for the Environmental and Earth Sciences*; Stoermer, E.F., Smol, J.P., Eds.; Cambridge University Press: Cambridge, UK, 2010.
12. Geraldes, A.M.; Boavida, M.J. Seasonal water level fluctuations: Implications for reservoir limnology and management. *Lakes Reserv. Res. Manag.* **2005**, *10*, 59–69. [CrossRef]
13. Vilhena, L.C.; Hillmer, I. The role of climate change in the occurrence of algal blooms: Lake Burragorang, Australia. *Limnol. Oceanogr.* **2010**, *55*, 1188–1200. [CrossRef]
14. Zhang, Q. The South-to-North Water Transfer Project of China: Environmental Implications and Monitoring Strategy. *J. Am. Water Resour. Assoc.* **2009**, *45*, 1238–1247. [CrossRef]
15. Chen, P.; Li, L.; Zhang, H. Spatio-temporal variability in the thermal regimes of the Danjiangkou Reservoir and its downstream river due to the large water diversion project system in central China. *Hydrol. Res.* **2016**, *47*, 104–127. [CrossRef]
16. Li, S.; Cheng, X.; Xu, Z.; Han, H.; Zhang, Q. Spatial and temporal patterns of the water quality in the the Danjiangkou Reservoir, China. *Hydrol. Sci. J.* **2009**, *54*, 124–134. [CrossRef]
17. Shen, Z.; Zhang, Q.; Yue, C.; Zhao, J.; Hu, Z.; Lv, N.; Tang, Y. The spatial pattern of land use/land cover in the water supplying area of the Middle-Route of the South-to-North Water Diversion (MR-SNWD) Project. *Acta Geogr. Sin.* **2006**, *6*, 9.
18. Zhu, M.; Tan, S.; Gu, S.; Zhang, Q. Characteristics of soil erodibility in the the Danjiangkou Reservoir Region, Hubei Province. *Chin. J. Soil Sci.* **2010**, *41*, 434–436.
19. Sicko-Goad, L.; Stoermer, E.F.; Ladewski, B.G. A morphometric method for correcting phytoplankton cell volume estimates. *Protoplasma* **1977**, *93*, 147–163. [CrossRef]
20. Rott, E. Some results from phytoplankton counting intercalibrations. *Schweiz. Z. Hydrol.* **1981**, *43*, 34–62. [CrossRef]
21. Hu, H.; Wei, Y. *The Freshwater Algae of China Systematics, Taxonomy and Ecology*; Sciences Press: Beijing, China, 2006.
22. Chinese Ministry of Environmental Protection (MEP). *Water Quality: Determination of Total Phosphorus—Ammonium Molybdate Spectrophotometric Method*; National Standard of China GB/T11893-1989; Chinese Ministry of Environmental Protection (MEP): Beijing, China, 1989.
23. Chinese Ministry of Environmental Protection (MEP). *Water Quality: Determination of Total Nitrogen—Alkaline Potassium Per-Sulfate Digestion Method*; National Standards of China HJ636-2012; Chinese Ministry of Environmental Protection (MEP) of China: Beijing, China, 2012.

24. Clarke, K.R. Non-parametric multivariate analyses of changes in community structure. *Aust. J. Ecol.* **1993**, *18*, 117–143. [CrossRef]

25. Oksanen, J.; Blanchet, F.G.; Kindt, R.; Legendre, P.; Minchin, P.R.; O'Hara, R.B.; Simpson, G.L.; Solymos, P.; Stevens, M.H.H.; Wagner, H. Vegan: Community Ecology Package. R Package Version 2.0-10. 2013. Available online: http://CRAN.R-project.org/package=vegan (accessed on 2 August 2017).

26. Anderson, M.J. A new method for non-parametric multivariate analysis of variance. *Austral Ecol.* **2001**, *26*, 32–46.

27. Levene, H. Robust tests for equality of variances. In *Contributions to Probability and Statistics*; Olkin, I., Ghurye, S.G., Hoeffding, W., Madow, W.G., Mann, H.B., Eds.; Stanford University Press: San Francisco, CA, USA, 1960.

28. R Development Core Team. R: A Language and Environment for Statistical Computing. 2008. Available online: http://www.R-project.org (accessed on 2 August 2017).

29. Wantzen, K.M.; Rothhaupt, K.O.; Mörtl, M.; Cantonati, M.; László, G.; Fischer, P. Ecological effects of water-level fluctuations in lakes: An urgent issue. *Hydrobiologia* **2008**, *613*, 1–4. [CrossRef]

30. Scharf, S.; Refilling, W. Ageing and water quality management of Brucher Reservoir. *Lakes Reserv. Res. Manag.* **2002**, *7*, 13–23. [CrossRef]

31. Mariani, M.A.; Padedda, B.M.; Kaštovský, J.; Buscarinu, P.; Sechi, N.; Virdis, T.; Lugliè, A. Effects of trophic status on microcystin production and the dominance of cyanobacteria in the phytoplankton assemblage of Mediterranean reservoirs. *Sci. Rep.* **2015**, *5*, 17964.

32. Kim, D.G.; Vargas, R.; Bond-Lamberty, B.; Turetsky, M.R. Effects of soil rewetting and thawing on soil gas fluxes: A review of current literature and suggestions for future research. *Biogeosciences* **2012**, *9*, 2459–2483. [CrossRef]

33. Meisner, A.; Bååth, E.; Rousk, J. Microbial growth responses upon rewetting soil dried for four days or one year. *Soil Biol. Biochem.* **2013**, *66*, 188–192. [CrossRef]

34. Gordon, H.; Haygarth, P.M.; Bardgett, R.D. Drying and rewetting effects on soil microbial community composition and nutrient leaching. *Soil Biol. Biochem.* **2008**, *40*, 302–311. [CrossRef]

35. Sommer, U.; Gliwicz, Z.M.; Lampert, W.; Duncan, A. The PEG-model of seasonal succession of planktonic events in fresh waters. *Arch. Hydrobiol.* **1986**, *106*, 433–471.

36. Reynolds, C.F. *The Ecology of Phytoplankton*; Cambridge University Press: Cambridge, UK, 2006.

37. Stevenson, R.J.; Bothwell, M.L.; Lowe, R.L. *Algal Ecology: Freshwater Benthic Ecosystem*; Academic Press: Manhattan, NY, USA, 1996.

38. Beer, S.V.; Spencer, W.; Bowes, G. Photosynthesis and growth of the filamentous blue-green alga *Lyngbya birgei* in relation to its environment. *J. Aquat. Plant Manag.* **1986**, *24*, 61–65.

39. Hudon, C.; De Sève, M.; Cattaneo, A. Increasing occurrence of the benthic filamentous cyanobacterium *Lyngbya wollei*: A symptom of freshwater ecosystem degradation. *Freshw. Sci.* **2014**, *33*, 606–618. [CrossRef]

40. Bridgeman, T.B.; Penamon, W.A. *Lyngbya wollei* in western Lake Erie. *J. Gt. Lakes Res.* **2010**, *36*, 167–171. [CrossRef]

41. Carmichael, W.W.; Evans, W.R.; Yin, Q.Q.; Bell, P.; Moczydlowski, E. Evidence for paralytic shellfish poisons in the freshwater cyanobacterium *Lyngbya wollei* (Farlow ex Gomont) comb. nov. *Appl. Environ. Microbiol.* **1997**, *63*, 3104–3110. [PubMed]

42. Gao, W.L.; Chen, Z.J.; Li, Y.Y.; Pan, Y.D.; Zhu, J.Y.; Guo, S.J.; Huang, J.; Hu, L.Q. Bioassessment of a drinking water reservoir using plankton: High throughput sequencing vs. traditionally morphological method. *Water* **2018**, *10*, 82. [CrossRef]

Article

Defining Seasonal Functional Traits of a Freshwater Zooplankton Community Using D^{13}C and D^{15}N Stable Isotope Analysis

Anna Visconti [1], Rossana Caroni [1,*], Ruth Rawcliffe [2], Amedeo Fadda [3], Roberta Piscia [1] and Marina Manca [1]

[1]   National Research Counsil, Water Research Institute, Largo Tonolli 50, 28922 Verbania, Italy;
      anninavis@hotmail.com (A.V.); r.piscia@ise.cnr.it (R.P.); m.manca@ise.cnr.it (M.M.)
[2]   Advanced Technology Institute, University of Surrey, Guildford GU2 7XH, UK; r.rawcliffe@surrey.ac.uk
[3]   University of Sassari, Department of Science for Nature and Environmental Resources, Piazza Università 21,
      07100 Sassari, Italy; amedeofadda@gmail.com
*   Correspondence: rossanarc@libero.it

Received: 30 November 2017; Accepted: 23 January 2018; Published: 27 January 2018

Abstract: Functional-based approaches are increasingly being used to define the functional diversity of aquatic ecosystems. In this study, we proposed the use of δ^{13}C and δ^{15}N stable isotopes as a proxy of zooplankton functional traits in Lake Maggiore, a large, deep subalpine Italian lake. We analyzed the seasonal pattern of δ^{13}C and δ^{15}N signatures of different crustacean zooplankton taxa to determine food sources, preferred habitats, and trophic positions of species throughout one year. The cladocerans *Daphnia longispina galeata* gr., *Diaphanosoma brachyurum*, and *Eubosmina longispina* were grouped into a primary consumer functional group from their δ^{13}C and δ^{15}N isotopic signatures, but while the former two species shared the same food sources, the latter exhibited a more selective feeding strategy. Cyclopoid copepods occupied a distinct functional group from the other secondary consumers, being the most ^{15}N enriched group in the lake. The δ^{15}N signature of calanoid copepods showed trophic enrichment in comparison to *Daphnia* and *Eubosmina* and linear mixing model results confirmed a predator-prey relationship. In our study, we have demonstrated that the use of δ^{13}C and δ^{15}N stable isotopes represented an effective tool to define ecological roles of freshwater zooplankton species and to determine functional diversity in a lake.

Keywords: functional diversity; zooplankton; seasonality; stable isotope analysis; trophic interactions

1. Introduction

Functional-based approaches are increasingly being used to study aquatic ecosystems as an alternative to traditional taxonomy-based approaches. Functional diversity is a biodiversity measure based on the ecological role of the species present in a community. Species-specific functional traits, or "what they do" [1], allows species to be defined by their interactions within an ecosystem [2] in terms of their ecological roles and how they interact with the environment and with other species [3].

Many recent ecological studies [2,4,5] suggest the importance of species ecological roles, and not just the number of taxonomic species, in the relationship between biodiversity and ecosystem functioning. This is a central concept if we are to understand and predict the resilience of a community to perturbations. If two species are deemed to be functionally alike and to occupy a similar trophic niche [6,7], the loss of one of those species is not likely to have an impact on the resource pool, as the other will increase its activity accordingly. The loss of functional diversity in the ecosystem is mitigated, as the species lost does not possess unique functional traits. Thus, the sum of organism functional traits within an ecosystem can be said to represent an indirect measure of its functional diversity [1,8].

Although the importance of functional diversity is widely recognized, there is no consensus on how to quantify functional diversity within a community, as relationships between the various indices have not yet been established [9].

Freshwater zooplankton play a key role in aquatic ecosystems in the transferal of biomass and energy from phytoplankton to top predators, e.g., [10,11]. While studies on phytoplankton have increasingly highlighted the importance of functional traits and functional classification in ecological studies, only a few have attempted this approach with zooplankton. Barnett et al. [2] applied the measure of functional diversity [12] to crustacean zooplankton communities. Quantitative functional traits considered were the C:N ratio, mean body size, and preferred food size range, while qualitative traits described the preferred habitat, food selectivity, and trophic position of each species.

Analysis of δ^{13}C and δ^{15}N stable isotopes is widely used to quantify food sources, trophic positions, and the interactions of organisms, as the δ^{13}C of a consumer can infer the assimilated source of dietary carbon [13,14] and δ^{15}N the trophic role [14,15]. Considering a functional-based perspective, δ^{13}C and δ^{15}N could identify ecological relations among taxa, such as competition and predation, ecological niche, habitat preference, and taxa redundancy, and help to define functional groups in an ecosystem. In this study, we propose the use of δ^{13}C and δ^{15}N stable isotope analysis to quantify some of the "qualitative functional traits" [2] for pelagic crustacean zooplankton taxa in Lake Maggiore, a large, deep subalpine lake in Italy. As species composition, diversity, and biomass of zooplankton can change significantly seasonally, especially in temperate lakes [16], the seasonal variation of δ^{13}C and δ^{15}N was used to define seasonal patterns in habitat preference, food sources, and taxa trophic position. This allowed an interpretation of taxa redundancy and a hypothesis of bottom-up and top-down mechanisms potentially driving the observed changes.

One use of stable isotopes is to determine the proportional contributions of several sources in a mixture. An example of a source proportion calculation includes the determination of various food sources in an animal's diet [17]. Linear mixing models are used to estimate proportions for two sources using isotopic signatures for a single element (e.g., δ^{13}C), or for three sources using isotopic signatures for two elements (e.g., δ^{13}C and δ^{15}N; [18]). In this study, we have used a linear mixing model [18,19] in order to discriminate the relative contribution of different preys in the diet of zooplankton consumers such as calanoid and cyclopoid copepods and the predatory cladoceran *Leptodora kindtii*. This further contributed to the understanding of food preference and taxa trophic position in the zooplankton community of Lake Maggiore.

2. Materials and Methods

Lake Maggiore (45°57′ N 8°32′ E 3°47′ W) is the second deepest (d_{max} 370 m) and largest (area 212.5 km^2, volume 37.5 km^3) subalpine lake in Italy. Being phosphorus-limited (TP$_{max}$ ca. 10 µg L^{-1}), the lake is oligotrophic and has recovered from eutrophication of the late 1970s [20,21].

Except for September, vertical zooplankton hauls from the surface to a depth of 50 m were collected. This follows the standard routine sampling for deep subalpine lakes, in which samples are collected within the upper 50 m depth, as previous research on the vertical distribution of zooplankton showed that this is the water layer in which zooplankton live [22]. Monthly samples were collected from April to November 2009, when total zooplankton biomass was ≥3 mg·m^{-3}, using a wide-mouth 450 µm mesh zooplankton net of diameter 0.58 m, filtering 13 m^3 lake water from three pelagic stations (G: 45°58′30″ N 8°39′09″ E, B: 45°54′28″ N 8°31′44″ E, L: 45°49′70″ N 8°34′70″ E) [16].

Zooplankton samples for the quantification of biomass were collected with a Clarke-Bumpus plankton sampler of a 126 µm mesh size and fixed in ethanol 96% to estimate the taxa-specific population density (ind·m^{-3}) and standing stock biomass (dry weight, mg·m^{-3}) [22]. Organisms for isotopic analyses were kept overnight in filtered (1.2 µm GF/C filters) lake water for gut clearance, before sorting into taxa and quantities suitable for isotopic analyses. The taxa analyzed were *Daphnia longispina galeata* gr., *Eubosmina longispina*, *Diaphanosoma brachyurum*, *Bythotrephes longimanus*,

Leptodora kindtii, adults of the calanoid copepods *Eudiaptomus padanus* and *Eudiaptomus gracilis*, and of the cyclopoid copepods *Mesocyclops leuckarti* and *Cyclops abyssorum*.

Samples were oven-dried for 24 h at 60 °C, before homogenizing and transferal into tin capsules of 5 × 9 mm in size. Depending on body mass, 50 to 700 individuals of each taxa were pooled to reach a minimum dry weight (DW) of 1 mg per sample. Three replicates of each taxa were run from each of the three sampling stations, as among-station differences were statistically non-significant ($p > 0.05$, Friedman Analysis of Variance, ANOVA test; [16]). The isotopic composition of organic carbon and nitrogen was determined from the analyses of CO_2 and N_2 by the G. G. Hatch Stable Isotope Laboratory at the University of Ottawa, Ontario, Canada, using a CE 1110 Elemental Analyser (Vario EL III manufactured by Elementar, Germany) and a DeltaPlus Advantage isotope ratio mass spectrometer (Delta XP Plus Advantage manufactured by Thermo, Bremen, Germany) coupled to a ConFlo III interface (Conflo II manufactured by Thermo, Bremen, Germany). The standard deviation of the analyses (SD) based on laboratory internal standards (C-55) was < 0.2‰ for both $\delta^{13}C$ and $\delta^{15}N$. Isotope ratios were expressed as the parts per thousand (‰) difference from a standard reference of PeeDee Belemnite for carbon and atmospheric N_2 for nitrogen:

$$\delta^{13}C, \delta^{15}N = \left[\left(\frac{R_{sample}}{R_{standard}}\right) - 1\right] \times 1000 \tag{1}$$

where R is the isotopic ratio: $^{13}C/^{12}C$ and $^{15}N/^{14}N$.

Lipids can be $\delta^{13}C$-depleted as a consequence of fractionation during lipid synthesis [23], which can lead to a misrepresentation of results as differences in predator-prey $\delta^{13}C$ could be greater than the expected 0.8‰ [24]. The C:N ratio was used as an indicator of lipid content. Invertebrates, including crustacean zooplankton, tend to have a C:N ratio of 4 [25], but C:N varies seasonally [26] and was as high as 7, so we used a revised version of the lipid normalizing procedure based on the C:N ratio [27], substituting the corrected parameters into Equations (2) and (3):

$$L = \frac{93}{1 + (0.246 \times (C \div N) - 0.75)^{-1}} \tag{2}$$

$$\delta^{13}C\prime = \delta^{13}C + D \times \left(I + \frac{390}{1 + \frac{287}{L}}\right) \tag{3}$$

where L is the proportion of lipid in the sample; C and N are the proportions of carbon and nitrogen in the sample, respectively; $\delta^{13}C\prime$ is the lipid normalized sample signature; $\delta^{13}C$ is the measured sample signature; D is the isotopic difference between the protein and lipid (7.018 ± 0.263); and I is a constant of 0.048 ± 0.013 [28].

Zooplankton $\delta^{13}C$ and $\delta^{15}N$ isotopic signatures were referred to that of the pelagic baseline, which was expressed by the primary consumer, *Daphnia longispina galeata* gr. This choice came from previous stable isotope studies in Lake Maggiore [16], showing that *Daphnia's* $\delta^{13}C$ signature in the different seasons was closely correlated with the signature of seston ($r = 0.86$; $p < 0.01$; N = 13), confirming that *Daphnia* was an appropriate proxy for the pelagic baseline against which the carbon isotopic signals of other zooplankton can be compared. $\Delta^{13}C$ was used to detect seasonal changes in taxa specific feeding behavior and assess the origin of carbon sources fueling the pelagic food web.

The carbon fractionation between consumer and resource (F = $\delta^{13}C_{cons} - \delta^{13}C_{diet}$) is ≤ 0.8‰ (± 1.1‰ S.D.) [24]. The $\delta^{15}N$ of consumers has been shown to be enriched 2.55‰ [29] for zooplankton, and was used to assess seasonal change in taxa-specific trophic position (T), as a consumer's carbon signature is related to the baseline (F $\leq 0.8 \pm 1.1$) by:

$$T = (E/\lambda) + 2 \tag{4}$$

where λ is the stepwise enrichment, E = 2.55‰ [29], and 2 is the value commonly assigned to the deviation of primary consumers from the pelagic isotopic baseline. A trophic level of T = 3 indicates that a consumer is feeding on a primary consumer, whereas T = 4 suggests that there is an intermediate prey.

When $\delta^{13}C$ of the predator lies between that of two different prey taxa, suggesting a simultaneous use of both sources, the percent carbon contribution (p; q) of each prey to the predator's diet was calculated by the 2-end member linear mixing model (2-em LMM), [18,19] as:

$$p = (\delta^{13}C_{predator} - \delta^{13}C_{prey2})/(\delta^{13}C_{prey1} - \delta^{13}C_{prey2}); q = 1 - p \tag{5}$$

where p and q are the relative contributions (%) of prey1 and prey2 carbon signatures to the predator $\delta^{13}C$ carbon signature ($\delta^{13}C_{predator}$).

When three potential prey sources were assessed, their isotopic signatures were partitioned by applying a 3-end member mixing model [18] to calculate the fractional contribution (p; q; z) of each of the three food sources to the predator's diet as:

$$p = ((\delta^{15}N_{prey3} - \delta^{15}N_{prey2})(\delta^{13}C_{predator} - \delta^{13}C_{prey2}) - (\delta^{13}C_{prey3}-\delta^{13}C_{prey2})(\delta^{15}N_{predator} - \delta^{15}N_{prey2}))/$$

$$((\delta^{15}N_{prey3} - \delta^{15}N_{prey2})(\delta^{13}C_{prey1} - \delta^{13}C_{prey2}) - (\delta^{13}C_{prey3} - \delta^{13}C_{prey2})(\delta^{15}N_{prey1} - \delta^{15}N_{prey2}));$$

$$q = ((\delta^{13}C_{predator} - \delta^{13}C_{prey3}) - (\delta^{13}C_{prey1} - \delta^{13}C_{prey3})p/(\delta^{13}C_{prey2} - \delta^{13}C_{prey3});$$

$$z = 1 - p - q \tag{6}$$

where p, q, and z are the relative carbon contribution (%) of prey (prey 1, 2, and 3). As required by the mixing model, $\delta^{13}C_{predator}$ and $\delta^{15}N_{predator}$ were corrected for trophic fractionation, by weighting the isotopic signature of the prey against their percentage contribution to total biomass on each sampling date.

The software IsoError 04 (https://www.epa.gov/eco-research/stable-isotope-mixing-models-estimating-source-proportion) [18] was used to perform all 3-end member LMM calculations. Statistical analyses (Shapiro-Wilkinson W-test, Spearman-Rank correlation, Hierarchical Cluster Analysis) were performed using the software Statistica 12 (version 12, TIBCO Statistica Company, Palo Alto, CA, USA) and Sigmaplot 11.0 (version 11, Systat Software Inc., San José, CA, USA).

Cluster analysis of the seasonal variation in $\delta^{13}C$ and $\delta^{15}N$ for each taxa was performed with the software Sigmaplot 11.0. An Euclidean distance measure was used as the data were continuous [30].

3. Results

3.1. Seasonal Variation in $\delta^{13}C$ and the Determination of Consumer Resources

The seasonal variation of $\delta^{13}C$ for the different zooplankton species is shown in Figure 1. The variation of $\delta^{13}C$ in *Daphnia* was most depleted in spring, with a value of −36.3‰ ± 0.6 (SD), and became most enriched during the summer in August, with a value of −26.0‰ ± 0.1 (SD). *Diaphanosoma* was present in the lake in October and November and its $\delta^{13}C$ signature overlapped with a strong correlation (p = 0.06; R = 0.66) with the $\delta^{13}C$ signature of *Daphnia*. Also, the $\delta^{13}C$ signature of *Bosmina* overlapped with that of *Daphnia* in June, July, and October. This similarity in $\delta^{13}C$ indicates that herbivorous cladocerans are sharing the same type of resources. When more $\delta^{13}C$-depleted values were recorded for *Bosmina* (−32.0‰ ± 0.1, SD) than for *Daphnia* (−29.7‰ ± 0.1, SD), this might suggest a deviation in dietary sources over the winter months (November).

The $\delta^{13}C$ signature of copepods was more $\delta^{13}C$-depleted than that of the pelagic baseline (annual mean of −26.4‰ ± 4.4, SD) all year round, with a $\delta^{13}C$ annual mean of −32.4‰ ± 4.6 (SD) for calanoid copepods and −34.5‰ ± 2.4 (SD) for cyclopoid copepods.

The procedure of lipid normalization for cyclopoid copepods decreased their average value of the $\delta^{13}C$ signature, but the relative values between species were unaffected. The seasonal trend of $\delta^{13}C$ signature for calanoid copepods generally traced the $\delta^{13}C$ pelagic baseline signature (F < 0.8). The $\delta^{13}C$

signature of *Bythotrephes* overlapped with the pelagic baseline (F < 0.8) from June to October, suggesting a tight dependence of this predatory cladoceran on pelagic food resources. However, in November, the δ^{13}C signature of *Bythotrephes* was less δ^{13}C -depleted (-26.8‰) and the least ^{13}C-depleted of all zooplankton taxa. The δ^{13}C signature of the other predatory cladoceran *Leptodora* was also related to the pelagic baseline (F < 0.8).

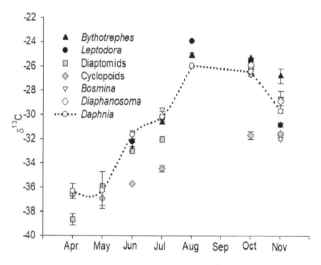

Figure 1. Seasonal changes in δ^{13}C signature (mean, $\pm$SE) in pelagic zooplankton taxa of Lake Maggiore in 2009. White symbols refer to primary consumers, grey and black to secondary consumers, with the dotted line referring to the pelagic isotopic baseline.

3.2. Seasonal Variation in δ^{15}N and Determination of Trophic Levels

The δ^{15}N seasonal pattern was the same for all zooplankton taxa, with more N-enriched values in early spring and late autumn, and less δ^{15}N-enriched values during the warm months (Figure 2). The seasonal variation in δ^{15}N was more pronounced in primary than in secondary consumers, as the predatory taxa *Bythotrephes*, *Leptodora*, and cyclopoid copepods had a small δ^{15}N range (NR) of 1.27‰, 1.32‰, and 1.74‰, respectively, whilst *Daphnia*, *Bosmina*, *Diaphanosoma*, and *Eudiaptomus* had a wider NR range of 3.75‰, 3.32‰, 3.57‰, and 2.50‰, respectively.

Cyclopoid copepods were the most δ^{15}N-enriched group, with δ^{15}N ranging between 7.92‰ in October and 9.67‰ in May, and with a F_{max} of 6.6‰.

On average, predatory cladocerans were more δ^{15}N-enriched than the pelagic isotopic baseline. The δ^{15}N signature of *Daphnia*, *Bosmina*, and *Diaphanosoma* overlapped ($\leq$5‰), with δ^{15}N of the latter two taxa significantly correlated with *Daphnia* (p < 0.001; Spearman Rank Correlation coefficient R > 0.9; N = 8, 12, respectively).

Calanoid copepods had enriched δ^{15}N signatures, ranging from 5.81‰ in October to 9.38‰ in April. Δ^{15}N enrichment with respect to the pelagic baseline varied between a maximum of 4.45‰ in October and a minimum of 2.65‰ in April, suggesting a change in trophic feeding level and differential exploitation of food resources.

Using mixing models to quantify the contribution of different prey to predators' diet, the contribution of the zooplankton preys assimilated by the consumer did not always match the zooplankton prey biomass present in the lake (Figure 3a,b). For example, in July, the estimated proportion of *Daphnia* and *Bosmina* in the diet of calanoid copepods was 53.2% and 46.8%, respectively, when *Daphnia* was present in the lake with a biomass of 80%. In October and November, the diet of calanoids copepods maintained a similar estimated contribution of *Daphnia* 57.7% and *Bosmina* 42.3%, when *Daphnia* was present in the lake with a biomass of 16.9% of total zooplankton biomass.

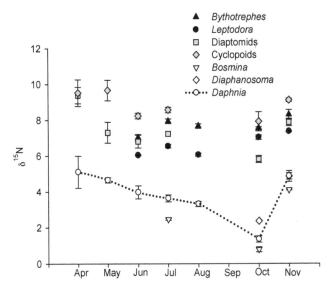

Figure 2. Seasonal changes in δ^{15}N signature (mean, ±SE) in pelagic zooplankton taxa of Lake Maggiore in 2009. White symbols refer to primary consumers, grey and black to secondary consumers, with the dotted line referring to the pelagic isotopic baseline.

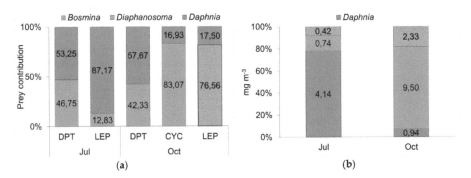

Figure 3. (a) Comparison between (a) contribution of different preys to the predators' diet calculated with the 2-em-LMM (JUL) and the 3-em-LMM (OCT-NOV); (b) biomasses (mg m^{-3}) of the potential preys in the sampling moment. Numbers within the bars correspond to calculated percentage values. DPT = calanoids; CYC = cyclopoids; LEP = *Leptodora*.

The diet of *Leptodora* in July was partitioned between *Daphnia* and *Bosmina* with 87.2% and 12.8%, respectively, while in autumn, the 3-linear mixing model indicated a higher consumption of *Diaphanosoma* (76.6%) than *Daphnia* (5.9%) and *Bosmina* (17.5%).

The diet of cyclopoid copepods in October and November was estimated by the mixing model to be 83.1% *Bosmina* and 16.9% *Daphnia*, when *Bosmina* was present in the lake, representing 75% of the total zooplankton biomass.

3.3. Determination of Functional Roles from $\delta^{13}C$ and $\delta^{15}N$

Cluster analysis of the seasonal variation in $\delta^{13}C$ for each taxon (Figure 4a) identified three major groups. The first split of the ordination separates the cyclopoid copepods from the other taxa, most likely as a result of the group utilizing deeper carbon sources in the pelagic zone. Lipid correction applied to $\delta^{13}C$ did not affect the seasonal pattern observed in the taxa. The second and further split in the cluster analysis grouped together the primary herbivorous cladocera *Daphnia longispina galeata* gr., *Diaphanosoma brachyurum*, and *Eubosmina longispina*, and the secondary consumers *Bythotrephes longimanus*, *Leptodora kindtii*, and calanoid copepods.

Cluster analysis of the seasonal variation in $\delta^{15}N$ for each taxon (Figure 4b) clearly grouped the taxa into two functional groups, the primary consumers, *D. longispina galeata* gr., *D. brachyurum*, and *E. longispina*, and the secondary consumers *B. longimanus*, *L. kindtii*, and copepods.

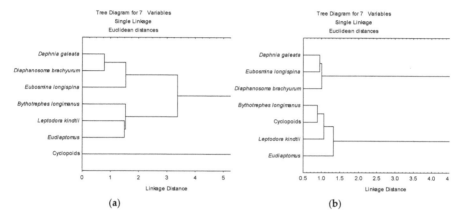

Figure 4. Cluster analysis of the seasonal variation of (**a**) $\delta^{13}C$ and (**b**) $\delta^{15}N$ for each zooplankton taxa using the Euclidean measure of distance.

4. Discussion

Functional biodiversity measures are based on the functional traits of the species in a community, rather than species richness, as it is interactions that determine the response of the ecosystem to a perturbation [31–33]. A methodology to determine the functional diversity within an ecosystem is quantifying the functional relationships between species as the distance measure of the branch length of the connecting functional dendrogram [12]. But it is not just the distance measure or clustering algorithm used that is important, it is the choice of which functional traits to include that is crucial [34].

In this study, we propose to use $\delta^{13}C$ signature as a proxy for determining habitat preference and foraging zone by inferring pelagic vs. littoral feeding preferences. We further analyzed the relationship between $\delta^{13}C$ taxa signature and phytoplankton succession during the studied year of 2009.

Confounding factors in a dynamic ecosystem like Lake Maggiore include seasonality and predation pressure variation. Seasonal changes in the littoral and pelagic $\delta^{13}C$ isotopic baseline were identified by a temporal shift towards less ^{13}C-depleted values in the summer, which is a trend commonly observed in thermally-stratified lakes [35,36]. In lacustrine systems, $\delta^{13}C$ and $\delta^{15}N$ of suspended particulate matter (seston) varies seasonally [37] because of differences in the allochthonous input, phytoplankton species composition, and primary productivity [38,39]. In Lake Maggiore, it has been demonstrated that $\delta^{13}C$ of the primary consumer *Daphnia longispina galeata* gr. tracks the isotopic composition of seston ($50 \leq$ size ≤ 126 μm) and of the pelagic baseline [16,35]. The choice of *Daphnia* as a useful isotopic baseline is supported by other studies [27], considering this taxa as a short-lived organism suited for fine scale temporal integration of pelagic $\delta^{13}C$ or $\delta^{15}N$ signatures.

If we consider the phytoplankton component of seston, there is variation in the fractionation of $\delta^{13}C$ between phytoplankton groups, with Chrysophyceae and Bacillariophyceae being more depleted in $\delta^{13}C$ than Cyanobacteria [37]. In Lake Maggiore, Bacillariophyceae are the most dominant phytoplankton group with a high biomass throughout the year [37,40]. In the year of our study (2009), a peak in the Bacillariophyceae biomass of *Aulacoseira* sp., *Asterionella* sp., and *Fragilaria* sp. was recorded in spring [37], when the cladoceran *Daphnia longispina galeata* gr. had the most $\delta^{13}C$-depleted values, while when Cyanobacteria biomass increased in the lake during the summer and autumn, *Daphnia longispina galeata* gr. had more $\delta^{13}C$-enriched values. This is suggestive of the opportunist nature of *Daphnia longispina galeata* gr., feeding on the most abundant component of the phytoplankton.

Rather than simply attributing organisms to different trophic groups *per se* from isolated values of $\delta^{15}N$, we analyzed the seasonal variation in $\delta^{13}C$ and $\delta^{15}N$ fractionation to identify changes in trophic interactions and feeding niche change. *Daphnia longispina galeata* gr., *Diaphanosoma brachyurum*, and *Eubosmina longispina* can be grouped into a "primary consumer" functional group from their $\delta^{13}C$ and $\delta^{15}N$ isotopic signatures and known trophic interactions. Determination of functional groups is important and consensus is being placed on the importance of how species are likely to react to a perturbation, which is particularly crucial in considering the consequences of species loss. In the cluster analysis we performed, the grouping of the $\delta^{13}C$ and $\delta^{15}N$ isotopic signature of the herbivorous cladocerans *Daphnia longispina galeata* gr. and *Diaphanosoma brachyurum*, indicates a use of the same seasonal carbon pool. This is also confirmed by the similar values of the two cladoceran taxa in their $\delta^{13}C$ signature. The position of *Eubosmina* in the cluster analysis might show a slightly different use of food resources, a hypothesis confirmed by its $\delta^{13}C$ depletion values observed in November. These results indicate a deviation in dietary sources and a possible shift towards a more selective feeding strategy for phytoplankton [29,41] over the winter, as phytoplankton tends to be more $\delta^{13}C$-depleted than detritus [28,41,42].

A redundant species can be generically defined as one that co-exists with an overlapping functional role and trophic niche to other species [5] in a community.

Bythotrephes is a cladoceran known to actively predate on *Daphnia* [43], which was indicated by its trophic $\delta^{15}N$ enrichment in comparison to *Daphnia*. However, in our study, when *Daphnia* was scarce in the lake in October and November, *Bythotrephes* seemed to switch its feeding preference to *Diaphanosoma* as its prey. This diet switching from *Daphnia* to *Diaphanosoma* during periods of *Daphnia* scarcity was also observed for *Leptodora*, as the $\delta^{15}N$ of *Leptodora* did not indicate that they fed on calanoid or cyclopoid copepods, rather that they preferentially exploit cladocerans.

Thus, the results in our study might indicate that *Daphnia longispina galeata* gr. and *Diaphanosoma brachyurum* are redundant species in the zooplankton community of Lake Maggiore, as their trophic roles are overlapping and interchangeable. Our hypothesis is further supported by the well-known interspecific competition of these two taxa for the same food sources in lakes [44].

The $\delta^{13}C$ signature of calanoid copepods in our study was more depleted than the *Daphnia* $\delta^{13}C$ signature. This has been observed in previous studies on freshwater zooplankton [28,38,41], and has been attributed to the calanoid copepods omnivorous diet of seston and small zooplankton like *Eubosmina* and rotifers [45–48].

The $\delta^{15}N$ isotopic signature of calanoid copepods showed a trophic stepwise enrichment in comparison to *Daphnia* and *Eubosmina*, indicative of a predator-prey relationship. This was confirmed by the results of the linear mixing model [18], showing that calanoid copepods preyed on *Daphnia* and *Eubosmina* in similar percentages in spring and in autumn. Although herbivorous during some life stages, copepods are one of the main groups of invertebrate predators in limnetic and littoral inland waters [45].

In our study, cyclopoid copepods occupied a distinct functional group from the other secondary consumers, being the most [15]N -enriched zooplankton group in the lake during the year. We recorded highly $\delta^{13}C$ depleted values for cyclopoid copepods, and a high fractionation with respect to the pelagic baseline (F ~ 7.6), suggesting that cyclopoids may be relying upon deeper carbon sources than those exploited by *Daphnia*. In deep lakes, the carbon isotopic signature is strongly influenced by depth, with organisms living in deeper layers characterized by more negative values than those living closer to the surface and in the littoral zone [42,49,50]. Moreover, in oligotrophic clear lakes, especially in periods of water column stratification, it is common to have a gradient of $\delta^{13}C$ POM (Particulate Organic Matter) with depth [51,52]. Matthew and Mazumder (2006) conducting a study in the oligotrophic Council Lake, recorded a $\delta^{13}C$ of POM decrease with depth in the water column, and zooplankton taxa at deeper depths also had a lower $\delta^{13}C$. Our hypothesis is reinforced by the cluster analysis with $\delta^{13}C$ values, which clearly show that cyclopoid copepods represented a distinct group, indicating a different utilization of carbon sources than other zooplankters, possibly found in deeper parts of the pelagic zone [49,53–55].

Because seasonal changes in the $\delta^{13}C$ isotopic baseline were identified by a temporal shift towards less ^{13}C depleted values in the littoral in the summer, this enabled the determination of the preferred foraging habitat of zooplankton taxa. The seasonal trend of cyclopoid copepods $\delta^{13}C$ values could also be related to the observed presence of the planktivorous fish roach (*Rutilus rutilus*) in the lake during the summer, as a predation-avoidance strategy by cyclopoid copepods from the increase in predator pressure in the pelagic. The $\delta^{13}C$ -enrichment observed in cyclopoid copepods in autumn may be explained by their return to the upper part of the pelagic zone when roach have migrated to the littoral zone to spawn [35].

The variation in $\delta^{15}N$ of *Bythotrephes* could also be related to the presence or absence of plankivorous fish in the pelagic zone, as they exhibit $\delta^{15}N$ depletion for the entirety of the period when planktivorous fish are feeding in the pelagic [35]. $\delta^{15}N$ enrichment increases when the predation-pressure decreases. As the variation of $\delta^{15}N$ between prey-predator decreases with niche occupation of increasing trophic levels, it is likely to be led by a top-down mechanism.

Cluster analysis with $\delta^{15}N$ seasonal variation clearly split the zooplankton taxa into two functional groups, the primary consumers, *Daphnia longispina galeata* gr., *Diaphanosoma brachyurum*, and *Eubosmina longispina*, and the secondary consumers, *Bythotrephes longimanus*, *Leptodora kindtii*, and the copepods. The secondary consumers were more ^{15}N -enriched than primary consumers. Cyclopoid copepods were feeding on the highest trophic level, which can be estimated to be 2–3 trophic levels higher than the primary producers, depending on whether a fractionation factor of 2.55 [29] or 3.4‰ [14] is used.

In this study, three or four trophic levels were identified in crustacean zooplankton of Lake Maggiore. In bio-magnification and energy and matter transfer in a pelagic food web, zooplankton is considered a crucial link between the primary producers and fish. In bio-energetic models, zooplankton are categorized as a "source" [56], or pooled into a singular grouping [57] of the trophic level of 2 [58,59], without taking into account species-specific traits or differences in life stage. Because the relative biomass of zooplankton taxa can significantly change during a calendar year [60], and because trophic positions of taxa and their relationships are dynamic during the year, we can conclude that freshwater zooplankton cannot be clustered together in the same ecological compartment. We propose the inclusion of seasonality and the dynamic of species trophic roles of zooplankton in the construction of models predicting bio-magnification capability.

In our study, we have demonstrated that the use of $\delta^{13}C$ and $\delta^{15}N$ stable isotope analysis represents an effective way to investigate the relationships present in the zooplankton community of a lake. In fact, the preferred habitat, food selectivity, and trophic position of each species could be defined. Through the use of $\delta^{13}C$ and $\delta^{15}N$ stable isotopes as a proxy of zooplankton functional traits, we can gain a better understanding of the ecological roles of the zooplankton species in the lake and thus define the functional diversity of the ecosystem. Moreover, combining the use of $\delta^{13}C$ and $\delta^{15}N$ stable isotopes with dietary analysis, we provided evidence that this could be an effective approach to infer functional groups, helping us understand the impact of functional differences in resource use. In particular, we have demonstrated that seasonal variations of $\delta^{13}C$ and $\delta^{15}N$ stable isotopes indicated a dynamic process of change in the relationships among zooplankton taxa, according to the different availability of food sources and of potential bottom-up and top-down (in particular fish predation) mechanisms present in the lake.

Acknowledgments: This study was funded by the CIPAIS (Commissione Internazionale per la Protezione delle Acque Italo-Svizzere) within the research program 'Research on the evolution of Lake Maggiore, Italy. Limnological studies. 2008–2012'. The authors would like to dedicate this paper to the memory of Giuseppe Morabito, limnologist and expert in freshwater phytoplankton ecology, who significantly contributed to the morpho-functional characterization of phytoplankton of subalpine deep lakes. He loved water, where he felt in perfect harmony and in which he found comfort and peace also during the course of his illness. We wish to thank three anonymous reviewers for critical comments that greatly contributed to improve the paper.

Author Contributions: M.M. conceived and designed the experiments; A.V., A.F., and R.P. performed the laboratory analysis; M.M. and A.V. analyzed the data; R.C. and A.V. wrote the paper with the contribution of other authors; R.R. helped with the Discussion.

Conflicts of Interest: The authors declare no conflict of interest.

References

1. Moss, B.R. *Ecology of Fresh Waters: Man and Medium, Past to Future*; John Wiley & Sons: Hoboken, NJ, USA, 2009.
2. Barnett, A.J.; Finlay, K.; Beisner, B.E. Functional diversity of crustacean zooplankton communities: Towards a trait-based classification. *Freshw. Biol.* **2007**, *52*, 796–813. [CrossRef]
3. Díaz, S.; Cabido, M. Vive la difference: Plant functional diversity matters to ecosystem processes. *Trends Ecol. Evol.* **2001**, *16*, 646–655. [CrossRef]
4. Tilman, D.; Knops, J.; Wedin, D.; Reich, P.; Richie, P.; Siemann, E. The influence of functional diversity and composition on ecosystem processes. *Science* **1997**, *277*, 1300–1302. [CrossRef]
5. Walker, B.; Kinzig, A.; Langridge, J. Plant attribute diversity, resilience, and ecosystem function: The nature and significance of dominant and minor species. *Ecosystems* **1999**, *2*, 95–113. [CrossRef]
6. Elton, C.S. *Animal Ecology*; University of Chicago Press: Chicago, IL, USA, 1927; p. 260.
7. Leibold, M.A. The niche concept revisited: Mechanistic models and community context. *Ecology* **1995**, *76*, 1371–1382. [CrossRef]
8. Petchey, O.L.; Gaston, K.J. Functional diversity: Back to basics and looking forward. *Ecol. Lett.* **2006**, *9*, 741–758. [CrossRef] [PubMed]
9. Mouchet, M.A.; Villeger, S.; Mason, N.W.; Mouillot, D. Functional diversity measures: An overview of their redundancy and their ability to discriminate community assembly rules. *Funct. Ecol.* **2010**, *24*, 867–876. [CrossRef]
10. Loseto, L.L.; Stern, G.A.; Ferguson, S.H. Size and biomagnification: How habitat selection explains beluga mercury levels. *Environ. Sci. Technol.* **2008**, *42*, 3982–3988. [CrossRef] [PubMed]
11. Forest, A.; Sampei, M.; Makabe, R.; Sasaki, H.; Barber, D.G.; Gratton, Y.; Fortier, L. The annual cycle of particulate organic carbon export in Franklin Bay (Canadian Artica): Environmental control and food web implications. *J. Geophys. Res. Oceans* **2008**, *113*, C03S05.
12. Petchey, O.L.; Gaston, K.J. Functional diversity (FD), species richness and community composition. *Ecol. Lett.* **2002**, *5*, 402–411. [CrossRef]
13. Peterson, B.J.; Fry, B. Stable isotopes in ecosystem studies. *Annu. Rev. Ecol. Syst.* **1987**, *18*, 293–320. [CrossRef]
14. Post, D.M. Using stable isotopes to estimate trophic position: Models, methods, and assumptions. *Ecology* **2002**, *83*, 703–718. [CrossRef]
15. Minagawa, M.; Wada, E. Stepwise enrichment of ^{15}N along food chains: Further evidence and the relation between δ^{15}N and animal age. *Geochim. Cosmochim. Acta* **1984**, *48*, 1135–1140. [CrossRef]
16. Visconti, A.; Manca, M. Seasonal changes in the δ^{13}C and δ^{15}N signatures of the Lago Maggiore pelagic food web. *J. Limnol.* **2011**, *70*, 263–271. [CrossRef]
17. Szepanski, M.M.; Ben-David, M.; Van Ballenberghe, V. Assessment of anadromous salmon resources in the diet of the Alexander Archipelago wolf using stable isotope analysis. *Oecologia* **1999**, *120*, 327–335. [CrossRef] [PubMed]
18. Phillips, D.L. Mixing models in analyses of diet using multiple stable isotopes: A critique. *Oecologia* **2001**, *127*, 166–170. [CrossRef] [PubMed]

19. Phillips, D.L.; Gregg, J.W. Uncertainty in source partitioning using stable isotopes. *Oecologia* **2001**, *127*, 171–179. [CrossRef] [PubMed]

20. De Bernardi, R.; Giussani, G.; Manca, M.; Ruggiu, D. Trophic status and the pelagic system in Lago Maggiore. *Hydrobiologia* **1990**, *191*, 1–8. [CrossRef]

21. Guilizzoni, P.; Marchetto, A.; Lami, A.; Gerli, S.; Musazzi, S. Use of sedimentary pigments to infer past phosphorus concentration in lakes. *J. Paleolimnol.* **2011**, *45*, 433–445. [CrossRef]

22. Manca, M.; Ruggiu, D. Consequences of pelagic food-web changes during a long-term lake oligotrophication process. *Limnol. Oceanogr.* **1998**, *43*, 1368–1373. [CrossRef]

23. McCutchan, J.H.; Lewis, W.M.; Kendall, C.; McGrath, C.C. Variation in trophic shift for stable isotope ratios of carbon, nitrogen, and sulfur. *Oikos* **2003**, *102*, 378–390. [CrossRef]

24. DeNiro, M.J.; Epstein, S. Influence of diet on the distribution of nitrogen isotopes in animals. *Geochim. Cosmochim. Acta* **1981**, *45*, 341–351. [CrossRef]

25. Matthews, B.; Mazumder, A. Temporal variation in body composition (C: N) helps explain seasonal patterns of zooplankton δ^{13}C. *Freshw. Biol.* **2005**, *50*, 502–515. [CrossRef]

26. McConnaughey, T.; McRoy, C.P. Food-web structure and the fractionation of carbon isotopes in the Bering Sea. *Mar. Biol.* **1979**, *53*, 257–262. [CrossRef]

27. Matthews, B.; Mazumder, A. Compositional and interlake variability of zooplankton effect baseline stable isotope signatures. *Limnol. Oceanogr.* **2003**, *48*, 1977–1987. [CrossRef]

28. Kiljunen, M.; Grey, J.; Sinisalo, T.; Harrod, C.; Immonen, H.; Jones, R.I. A revised model for lipid-normalizing δ^{13}C values from aquatic organisms, with implications for isotope mixing models. *J. Appl. Ecol.* **2006**, *43*, 1213–1222. [CrossRef]

29. Petchey, O.L.; Gaston, K.J. Dendrograms and measuring functional diversity. *Oikos* **2007**, *116*, 1422–1426. [CrossRef]

30. Petchey, O.L.; Hector, A.; Gaston, K.J. How do different measures of functional diversity perform? *Ecology* **2004**, *85*, 847–857. [CrossRef]

31. McCann, K.S. The diversity–stability debate. *Nature* **2000**, *405*, 228–233. [CrossRef] [PubMed]

32. Woodward, G.; Papantoniou, G.; Edwards, F.; Lauridsen, R.B. Trophic trickles and cascades in a complex food web: Impacts of a keystone predator on stream community structure and ecosystem processes. *Oikos* **2008**, *117*, 683–692. [CrossRef]

33. Petchey, O.L.; O'Gorman, E.J.; Flynn, D.F. A functional guide to functional diversity measures. In *Biodiversity, Ecosystem Functioning, and Human Wellbeing. An Ecological and Economic Perspective*; Oxford University Press: Oxford, UK, 2009; pp. 49–60. ISBN 9780199547951.

34. Visconti, A.; Volta, P.; Fadda, A.; Di Guardo, A.; Manca, M. Seasonality, littoral versus pelagic carbon sources, and stepwise 15N-enrichment of pelagic food web in a deep subalpine lake: the role of planktivorous fish. *Can. J. Fish. Aquat. Sci.* **2013**, *71*, 436–446. [CrossRef]

35. Poma, G.; Volta, P.; Roscioli, C.; Bettinetti, R.; Guzzella, L. Concentrations and trophic interactions of novel brominated flame retardants, HBCD, and PBDEs in zooplankton and fish from Lake Maggiore (Northern Italy). *Sci. Total Environ.* **2014**, *481*, 401–408. [CrossRef] [PubMed]

36. Lehmann, M.F.; Bernasconi, S.M.; Barbieri, A.; Simona, M.; McKenzie, J.A. Interannual variation of the isotopic composition of sedimenting organic carbon and nitrogen in Lake Lugano: A long-term sediment trap study. *Limnol. Oceanogr.* **2004**, *49*, 839–849. [CrossRef]

37. Caroni, R.; Free, G.; Visconti, A.; Manca, M. Phytoplankton functional traits and seston stable isotopes signature: A functional-based approach in a deep, subalpine lake, Lake Maggiore (N. Italy). *J. Limnol.* **2012**, *71*, 84–94. [CrossRef]

38. Grey, J.; Jones, R.I.; Sleep, D. Seasonal changes in the importance of the source of organic matter to the diet of zooplankton in Loch Ness, as indicated by stable isotope analysis. *Limnol. Oceanogr.* **2001**, *46*, 505–513. [CrossRef]

39. Vuorio, K.; Meili, M.; Sarvala, J. Taxon-specific variation in the stable isotopic signatures (δ^{13}C and δ^{15}N) of lake phytoplankton. *Freshw. Biol.* **2006**, *51*, 807–822. [CrossRef]

40. Morabito, G.; Sili, C.; Panzani, P.; Oggioni, A. Dinamica stagionale e distribuzione orizzontale di fitoplancton, carbonio organico e batterio plankton: Struttura dei popolamenti fitoplanctonici. In *Ricerche Sull'evoluzione del Lago Maggiore. Aspetti Limnologici. Programma Quinquennale 2008–2012*; Commissione Internazionale per la Protezione delle Acque Italo-Svizzere, 2009; pp. 53–66. ISSN 1013-8099. Available online: http://www.cipais.org/pdf/Limno_Maggiore_-_Rapporto_2009.pdf (accessed on 26 January 2018).

41. Grey, J.; Jones, R.I.; Sleep, D. Stable isotope analysis of the origins of zooplankton carbon in lakes of differing trophic state. *Oecologia* **2000**, *123*, 232–240. [CrossRef] [PubMed]

42. Jones, R.I.; Grey, J.; Sleep, D.; Quarmby, C. An assessment, using stable isotopes, of the importance of allochthonous organic carbon sources to the pelagic food web in Loch Ness. *Proc. R. Soc. Lond. B Biol. Sci.* **1998**, *265*, 105–110. [CrossRef]

43. Manca, M.; Vijverberg, J.; Polishchuk, L.; Voronov, D.A. *Daphnia* body size and population dynamics under predation by invertebrate and fish predators in Lago Maggiore: An approach based on contribution analysis. *J. Limnol.* **2008**, *67*, 15–21. [CrossRef]

44. Matveev, V.F. Effect of Competition on the Demography of Planktonic Cladocerans: *Daphnia* and *Diaphanosoma*. *Oecologia* **1987**, *74*, 468–477. [CrossRef] [PubMed]

45. Brandl, Z. Freshwater copepods and rotifers: Predators and their prey. *Hydrobiologia* **2005**, *546*, 475–489. [CrossRef]

46. Williamson, C.E.; Butler, N.M. Predation on rotifers by the suspension-feeding calanoid copepod *Diaptomus pallidus*. *Limnol. Oceanogr.* **1986**, *31*, 393–402. [CrossRef]

47. Kling, G.W.; Fry, B.; O'Brien, W.J. Stable isotopes and planktonic trophic structure in arctic lakes. *Ecology* **1992**, *73*, 561–566. [CrossRef]

48. Luecke, C.; O'Brien, W.J. The effect of *Heterocope* predation on zooplankton communities in arctic ponds. *Limnol. Oceanogr.* **1993**, *28*, 367–377. [CrossRef]

49. Vander Zanden, M.J.; Rasmussen, J.B. Primary consumer δ^{13}C and δ^{15}N and the trophic position of aquatic consumers. *Ecology* **1999**, *80*, 1395–1404. [CrossRef]

50. Vander Zanden, M.J.; Chandra, S.; Park, S.K.; Vadeboncoeur, Y.; Goldman, C.R. Efficiencies of benthic and pelagic trophic pathways in a subalpine lake. *Can. J. Fish. Aquat. Sci.* **2006**, *63*, 2608–2620. [CrossRef]

51. Cattaneo, A.; Manca, M.; Rasmussen, J.B. Peculiarities in the stable isotope composition of organisms from an alpine lake. *Aquat. Sci.* **2004**, *66*, 440–445. [CrossRef]

52. Post, D.M.; Pace, M.L.; Hairston, N.G. Ecosystem size determines food-chain length in lakes. *Nature* **2002**, *405*, 1047–1049. [CrossRef] [PubMed]

53. Quay, P.D.; Emerson, S.R.; Quay, B.M.; Devol, A.H. The carbon cycle for Lake Washington—A stable isotope study. *Limnol. Oceanogr.* **1986**, *31*, 596–611. [CrossRef]

54. Del Giorgio, P.A.; France, R.L. Ecosystem-specific patterns in the relationship between zooplankton and POM or microplankton δ^{13}C. *Limnol. Oceanogr.* **1996**, *41*, 359–365. [CrossRef]

55. Jeppesen, E.; Søndergaard, M.; Christoffersen, K.; Theil-Nielsen, J.; Jürgens, K. Cascading trophic interactions in the littoral zone: An enclosure experiment in shallow Lake Stigsholm, Denmark. *Arch. Hydrobiol.* **2002**, *153*, 533–555. [CrossRef]

56. Barwick, M.; Maher, W. Biotransference and biomagnification of selenium copper, cadmium, zinc, arsenic and lead in a temperate seagrass ecosystem from Lake Macquarie Estuary, NSW, Australia. *Mar. Environ. Res.* **2003**, *56*, 471–502. [CrossRef]

57. Paterson, M.J.; Rudd, J.W.; St. Louis, V. Increases in Total and Methylmercury in Zooplankton following Flooding of a Peatland Reservoir. *Environ. Sci. Technol.* **1998**, *32*, 3868–3874. [CrossRef]

58. Connolly, J.P.; Pedersen, C.J. A thermodynamic-based evaluation of organic chemical accumulation in aquatic organisms. *Environ. Sci. Technol.* **1988**, *22*, 99–103. [CrossRef] [PubMed]

59. Campfens, J.; Mackay, D. Fugacity-based model of PCB bioaccumulation in complex aquatic food webs. *Environ. Sci. Technol.* **1997**, *31*, 577–583. [CrossRef]

60. Morabito, G.; Mazzocchi, M.G.; Salmaso, N.; Zingone, A.; Bergami, C.; Flaim, G.; Accoroni, S.; Basset, A.; Bastianini, M.; Belmonte, G.; et al. Plankton dynamics across the freshwater, transitional and marine research sites of the LTER-Italy Network. Patterns, fluctuations, drivers. *Sci. Total Environ.* **2017**, submitted.

Article

Carbon and Nitrogen Isotopic Signatures of Zooplankton Taxa in Five Small Subalpine Lakes along a Trophic Gradient

Roberta Piscia [1,*], Emanuela Boggio [1,2], Roberta Bettinetti [1,2], Michela Mazzoni [2] and Marina Manca [1]

[1] National Research Council, Water Research Institute, CNR ISE Largo Tonolli 50, 28922 Verbania, Italy; manu.intra@yahoo.it (E.B.); roberta.bettinetti@uninsubria.it (R.B.); m.manca@ise.cnr.it (M.M.)

[2] DiSTA, University of Insubria, Via Valleggio 11, 22100 Como, Italy; mmazzoni@studenti.uninsubria.it

* Correspondence: r.piscia@ise.cnr.it; Tel.: +39-0323-518-333

Received: 9 October 2017; Accepted: 17 January 2018; Published: 22 January 2018

Abstract: Interest in Stable Isotopes Analyses (SIA) is increasing in freshwater ecology to better clarify ecosystems' functioning. By measuring carbon and nitrogen isotopic signatures, food sources and organism trophic levels can be tracked, providing quantitative estimates of bi-dimensional niches. In order to describe some general patterns of carbon and nitrogen stable isotope signatures in lakes, we applied SIA to zooplankton community in five subalpine lakes sampled in spring and summer along a trophic gradient (from oligotrophy to hypereutrophy). Within zooplankton taxa, temporal variation in food sources and trophic levels were compared to find out taxon-specific patterns. Carbon and nitrogen isotopic signatures differed among the five lakes, reflecting depth, topography, and trophic status of the lakes. Carbon isotopic signatures varied more considerably in deeper and larger lakes (Mergozzo and Pusiano) than in a shallower and smaller lake (Lake Endine). Nitrogen isotopic signatures were generally more enriched in lakes Pusiano and Moro than in Lake Mergozzo, whereas in summer, they were depleted in all lakes. These observations indicate that zooplankton taxa specific trophic roles differed among lakes and in time.

Keywords: Stable Isotopes Analysis; trophic gradient; small lakes; zooplankton

1. Introduction

Stable isotopes are increasingly used in aquatic ecological studies to clarify food web functioning by quantifying carbon and nitrogen flows through water ecosystems [1]. The great advantage of this approach is that complex interactions among organisms are simultaneously captured [2–4], providing information about roles of the different taxa in the environment.

The basic idea is that the isotope ratio of a consumer depends on its diet. Stable carbon isotope ratio ($\delta^{13}C‰$) reflects the input of carbon revealing the contributions of different food sources, and nitrogen isotope ($\delta^{15}N‰$) indicates the trophic role because a consumer is typically enriched with respect to its diet (e.g., [5–8]).

Increasing $\delta^{15}N$ enrichment is usually observed with the increasing lake trophic status [9–11]. Carbon contents of planktonic grazers and their food sources differ in response to trophic status in lakes, mainly because of differences in contribution of phytoplankton to the food sources [10]. On the other hand, less to more ^{13}C-depleted carbon signatures from littoral to pelagic carbon sources are observed (e.g., [6,12]).

Lake ecosystem functioning is influenced by length and complexity of food webs, which in turn vary with environmental conditions. In this context, zooplankton plays a crucial role, linking primary producers to secondary consumers (e.g., zooplankton predators and fish) in a dynamic process, affecting transfer of matter and energy through lake ecosystems.

Complexity of trophic interactions increases along with lake size: increasing area and depth allow for increasing taxa and competitors, with a higher degree of specialization of predators to prey [13].

In the present study, we compared carbon and nitrogen isotopic signatures of zooplankton taxa in five minor subalpine Italian lakes with different trophic status to outline some patterns already observed in other natural and artificial lakes. We tested if the baseline and the role of different zooplankton taxa vary depending on lake trophic status and size and along seasons.

2. Materials and Methods

2.1. Study Sites

Five Southern subalpine lakes located within the River Po catchment basin (Northern Italy) at an altitude between 194 and 389 m a.s.l. (Figure 1 and Table 1) were selected as representative of different trophic status (from the hypereutrophic lakes Pusiano and Comabbio to the oligotrophic Lake Mergozzo) and size (from the small and shallow Lake Comabbio to the deepest and relatively large Lake Mergozzo).

All lakes, except Lake Moro, are located in anthropized areas, and the vegetation of all catchment basin is composed by permanent meadows, softwood, and hardwood [14,15].

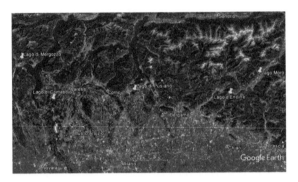

Figure 1. Satellite view of studied lakes (from Google Earth).

Table 1. Morpho-edaphic characteristics of studied lakes [16–19] (P_{tot}: total phosphorus).

Lake	Altitude (m a.s.l.)	Geographic Coordinates (Latitude Longitude)	Mixing Regime	Volume (10^6 m^3)	Depth$_{max}$ (m)	P_{tot} ($\mu g\ L^{-1}$)
Mergozzo	194	45°57′20″ N 8°27′52″ E	monomictic	83	73	1
Moro	389	45°52′47″ N 10°09′25″ E	meromictic	4	42	8
Endine	334	45°46′41″ N 9°56′22″ E	dimictic	12	9	17
Comabbio	243	45°45′55″ N 8°41′37″ E	polymictic	16	8	72
Pusiano	259	45°48′09″ N 9°16′17″ E	monomictic	69	24	74

The principal morpho-edaphic characteristic of the five lakes are reported in Table 1. All lakes, except Lake Mergozzo, may be covered with ice during very cold winters. As in most temperate lakes, the zooplankton community of these lakes is composed of cold and warm stenothermal and eurythermal taxa (Table S1).

2.2. Zooplankton Sampling and SIA

Zooplankton samples were collected at the deepest part of the lake during growing season in spring (May) and late summer (September) 2004 by vertical hauls with a 126 µm light plankton nylon net of 20 cm-diameter opening mouth and preserved in ethanol 95%. Crustacean zooplankton were sorted by taxa in two replicates, oven-dried (60 °C for 48 h), finely ground, and transferred into tin capsules. Because of different feeding habits of copepod developmental stages, we performed SIA only on samples made up of adults and/or sub-adults of cyclopoids and diaptomids, respectively. Depending on individual body weight, between 50 and 600 specimens were necessary to get 1 mg d.w./replicate sample. Samples were sent to the G.G. Hatch Stable Isotope Laboratory of Ottawa (Ottawa, ON, Canada) for SIA analyses (Continuous Flow Isotope-Ratio Mass Spectrometry for $\delta^{13}C$, $\delta^{15}N$). Data were reported in delta (δ) notation, the units expressed as ‰ and defined as $\delta = ((R_x - R_{std})/R_{std}) \times 1000$, where R is the ratio of the abundance of the heavy to the light isotope, x denotes sample and std is the abbreviation for standard. All $\delta^{15}N$ was reported as ‰ vs. AIR and normalized to internal standards calibrated to International standards IAEA-N1 (+0.4‰), IAEA-N$_2$ (+20.3‰), USGS-40 (−4.52‰), and USGS-41 (47.57‰). All $\delta^{13}C$ was reported as ‰ vs. V-PDB and normalized to internal standards calibrated to International standards IAEA-CH-6 (−10.4‰), NBS-22 (−29.91‰), USGS-40 (−26.24‰), and USGS-41 (37.76‰).

On each date, *Daphnia* signatures identified lake-specific pelagic baseline, representing time-specific signature of seston particles (in a range 1.2–50 µm), fueling the pelagic food web [5,6,8,12,20–22]. Taxa-specific signatures were referred to time-specific baseline signatures, allowing for tracing seasonal changes of deeper vs. more surficial carbon sources and relative position in the food web.

Carbon isotopic distance with respect to baseline (ID = $\delta^{13}C_{consumer} - \delta^{13}C_{baseline}$) was used to identify pelagic carbon sources (ID $\leq$ 0.8 $\pm$ 1.1) [2]. We calculated taxa-specific enrichment (E = $\delta^{15}N_{consumer} - \delta^{15}N_{baseline}$) to infer trophic position of each taxa with respect to the baseline.

3. Results

Carbon and nitrogen isotopic signatures of baseline were lake and season-specific (Figure 2). In spring, $\delta^{13}C$ ranged between high depleted values (ca. −36 $\delta^{13}C$‰ in lakes Pusiano and Moro) and the least depleted value of Lake Comabbio (−20 $\delta^{13}C$‰). In summer, $\delta^{13}C$ signatures were less ^{13}C depleted than in spring in three out of five lakes, namely, Mergozzo, Pusiano, and Moro. An opposite trend characterized the shallowest Lake Comabbio, with a more ^{13}C depleted carbon signature in summer than in spring. The lowest seasonal range of variation (of ca. 2 $\delta^{13}C$‰) was measured in Lake Endine. In spring, more enriched baseline $\delta^{15}N$ signatures characterized lakes Pusiano and Moro, while Lake Mergozzo was the least enriched. In summer, nitrogen isotopic signatures were less enriched than in spring in all lakes but Comabbio, in which no seasonal variation was measured. Despite of the clear difference in the isotopic fingerprint of the baselines among lakes, we did not find statistically significant correlations between them and morpho-edaphic characteristics.

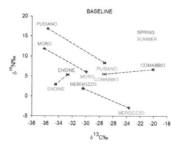

Figure 2. Carbon and nitrogen isotopic fingerprint of the pelagic baselines (*Daphnia*) in spring (green) and summer (red) in all lakes.

In spring, cyclopoids and diaptomids were the most enriched with respect to the baseline (Figures 3a and 4a,b). Marked differences were found among lakes δ^{15}N signatures, while the range of variation in δ^{13}C was relatively small.

Lakes Pusiano and Moro lay on the top of the trophochemical graph, the former also characterized by an almost complete overlapping of isotopic signatures of cyclopoids and diaptomids; in the latter, cyclopoids were more enriched than diaptomids with respect to the baseline (respectively 5.47 δ^{15}N‰ and 3.28 δ^{15}N‰). In both lakes Pusiano and Moro, copepods were tightly linked with pelagic carbon sources. In Lake Mergozzo, carbon sources of cyclopoids were clearly separated from the pelagic isotopic signature, being more δ^{13}C depleted than the baseline. In this lake, as in all others, cyclopoids had the highest δ^{15}N‰ in spring. Diaptomids were closely related to the cyclopoids for nitrogen isotopic signature, and the enrichment with respect to the pelagic baseline was high (5.70 δ^{15}N‰). In Lake Endine, diaptomids and cyclopoids shared similar δ^{15}N signatures, while being well separated for δ^{13}C signature: cyclopoids relying upon pelagic carbon sources, and diaptomids being more δ^{13}C depleted than pelagic baseline carbon signature. In Lake Endine, diaptomids appeared tightly related to *Bosmina*, with respect to which they were δ^{15}N‰ enriched of ca. 5. Enrichment of the cyclopoids with respect to *Daphnia* was also high (4.06). Carbon fractionation between diaptomids and the pelagic baseline is indicative of deeper food sources than those integrated by *Daphnia* and better represented by *Bosmina*. The most distinct isotopic signatures were those of Lake Comabbio, in which carbon signatures of diaptomids and of the pelagic baseline were the least depleted, while being δ^{15}N enrichment substantial (4.93), although lower than in Lake Mergozzo (5.70).

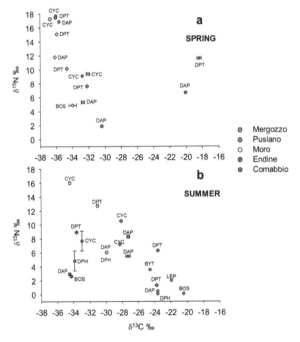

Figure 3. Trophochemical graph of zooplankton taxa signatures in spring (**a**) and summer (**b**). Error bars represent the Standard Error. CYC: cyclopoids; DPT: diaptomids; DAP: *Daphnia*; BOS: *Bosmina*; DPH: *Diaphanosoma*; BYT: *Bythotrephes*; LEP: *Leptodora*.

Differences among the lakes were clearer in summer, when the range of variation in carbon signatures was larger than in spring, ranging between the least depleted taxa signatures from

Lake Mergozzo to the most depleted taxa signatures of Lake Endine (Figure 3b). In the latter, *Diaphanosoma* signatures fully overlapped the pelagic baseline. Diaptomids also relied on pelagic food sources, although thanks to their nitrogen isotopic signature, they were the most enriched with respect to the baseline (6.70; Figure 4a,b).

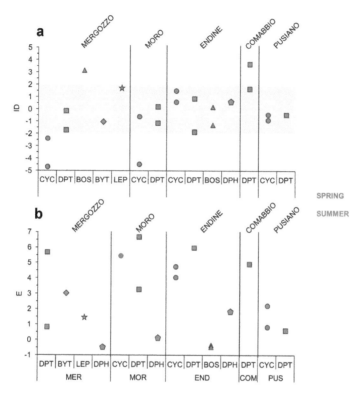

Figure 4. (a) Carbon isotopic distance (ID) of zooplankton taxa with respect to the baseline (*Daphnia*) in the studied lakes. ID values ranging between −1.9 and +1.9 are highlighted by the grey bar; (b) Nitrogen enrichment (E) of zooplankton taxa relying upon lake pelagic baseline (*Daphnia*). CYC: cyclopoids; DPT: diaptomids; BOS: *Bosmina*; DPH: *Diaphanosoma*; BYT: *Bythotrephes*; LEP: *Leptodora*.

A full overlapping between *Daphnia* and *Bosmina* and a close similarity with *Diaphanosoma* characterized Lake Endine's summer signatures, where diaptomids were the most δ^{15}N-enriched. In this lake, between-season variation was the least for both carbon and nitrogen signatures. Intermediate carbon signatures characterized lakes Moro and Comabbio. Lake Mergozzo's zooplankton signatures were on the right and bottom part of the trophochemical graph. In this lake, *Diaphanosoma* was clearly separated from *Bosmina* and from the *Daphnia* pelagic baseline, being characterized by the least δ^{13}C-depleted signature. *Leptodora* was δ^{15}N enriched with respect to *Diaphasomoma*, sharing a carbon signature intermediate between those of *Daphnia*, *Diaphanosoma*, and *Bosmina*. On the other side, isotopic signature of *Bythotrephes* was consistent with those of *Daphnia* (representing the pelagic carbon baseline), with δ^{15}N-enrichment of 3.08. In this lake, the cyclopoids were clearly separated from all other taxa and from the pelagic baseline, being carbon signature highly depleted (δ^{13}C = −28.30) and nitrogen signature highly enriched (δ^{15}N = 7.24).

Changes in time in C isotopic values of the baseline were substantial, ranging between most ^{13}C-depleted values in spring to least depleted values in summer (extremes being −36.74‰ of cyclopoids in Lake Moro in spring and −20.46 of *Bosmina* in Lake Mergozzo in summer; Table S2). The least $\delta^{13}C$-depleted signature was measured in Lake Comabbio, in which a large variability in $\delta^{15}N$ signature was measured in the diaptomids. Least depleted summer carbon isotopic signatures were measured in all lakes but Endine, where temporal variations of all taxa were very small. In Lake Mergozzo, within time variation in diaptomids' carbon isotopic signature was wider than that of *Daphnia*, but as in most lakes, variability in $\delta^{15}N$ was very small.

4. Discussion

Carbon and nitrogen stable isotope signatures are lake-specific. In deeper lakes, more depleted carbon signatures are expected than in shallower ones; lakes with a higher trophic status are usually more enriched in $\delta^{15}N$ than those with a lower trophic status [9–11].

Our dataset spanned from the shallowest Lake Comabbio to the deepest Lake Mergozzo, also including the small and relatively deep Lake Moro, which is characterized by cold waters. The dataset also comprised the oligotrophic Lake Mergozzo and the hypereutrophic Lake Pusiano, thus covering a wide range of lake trophic status. Our results confirm some general patterns. In the deep subalpine Lake Mergozzo, seasonal changes in taxa-specific signatures were consistent with those observed in other deep subalpine lakes, such as Lake Maggiore [12]. With summer thermal stratification, carbon signature of the pelagic baseline tends towards less ^{13}C depleted values. This observation probably reflects changes in phytoplankton isotopic signature [21]. Phytoplankton, during growth season, exhibits $\delta^{13}C$ values less negative, because of reduced isotopic carbon fractionation at high cell densities and/or a shift on exploitation of HCO_3^- as carbon source instead of CO_2 [21–31]. *Daphnia*, being an unselective filter-feeder integrates seston particle up to a size fitting the intersetular distance of filtering combs, in turn depending on *Daphnia* body size. In deep subalpine lakes, seston ≤50 μm is mainly composed by phytoplankton algal cells [32]. Carbon fractionation with respect to seston is therefore negligible, and seasonal changes in *Daphnia* carbon signature entirely overlap those of phytoplankton, thus providing a reliable pelagic baseline [6,8,33–35]. Changes in carbon baseline signature with thermal stratification in turn reflect changes in taxa composition and carbon dynamics of phytoplankton in the upper water layers [6,20,21].

According to this interpretation, also in lakes Pusiano and Moro, summer thermal stratification led to increased contribution of the upper layers to phytoplankton production in summer. In Lake Comabbio, and to a lesser extent in Lake Endine, such change is not confirmed, the summer carbon signature of both *Daphnia* pelagic baseline and the diaptomids being more $\delta^{13}C$ depleted than in spring. Such opposite pattern is consistent with a recruitment of *Daphnia cucullata* and of *Eudiaptomus padanus* from the lake littoral, characterized by less negative carbon values (e.g., [36]), into the lake profundal during spring. Horizontal migration of zooplankton has been well documented in a number of cases, such as *Eudiaptomus padanus* in the deep nearby Lake Maggiore [36].

Very close or overlapping carbon and nitrogen signatures of primary consumers, such as *Bosmina* and *Diaphanosoma*, suggest the exploitation of the same food sources in Lake Endine, particularly in summer. In Lake Mergozzo, isotopic signatures suggest that *Diaphanosoma*, *Daphnia*, and *Eudiaptomus* have a very similar trophic role, with *Bosmina* being well separated from the other three zooplankton taxa, clearly relying upon more superficial (or more littoral) carbon sources. The least $\delta^{13}C$-depleted signature, in fact, is indicative of warmer waters or littoral sources. Spatial and temporal variations in carbon and nitrogen isotopic signatures in different compartments of a single lake were assessed by former studies [37,38]. Previous research on Lake Maggiore demonstrated that littoral carbon isotopic signature follows the same pattern as pelagic carbon with increasingly less depleted ^{13}C values in summer than in spring, the shift being responsible for a clear separation of littoral with respect to pelagic isotopic fingerprint [6,12].

Previous work on Lake Maggiore [6,12] suggested a relatively narrow range of carbon isotopic signature of cyclopoids, along with a clear separation from summer pelagic signature. The two peculiar traits are interpreted as a clear specificity of habitat and food selection [37]. Our dataset only partially confirms these two traits: while separation with respect to summer pelagic carbon is confirmed in lakes Moro and Mergozzo, such is not the case in lakes Pusiano and Endine, where cyclopoids do not show differences with respect to pelagic carbon signature in summer. Seasonal variability of carbon signature was small in lakes Endine and Moro, while being very large, almost comparable to that of the unselective filter-feeder *Daphnia*, in the other lakes. Different taxa may contribute along the season to the cyclopoids: in Lake Pusiano, where variability was the largest, three species have been reported, namely *Cyclops abyssorum*, *Mesocyclops leuckarti*, and *Thermocyclops hyalinus*. Previous studies have suggested interspecific differences of cyclopoid copepod species in the exploitation of food sources (e.g., [38,39]) and their relative trophic position [37]. Moreover, while *Cyclops* is known to grow and reach maximum density during cold months, the other two species tend to increase during the warm season. Such a strong seasonality explains the wide range of variability in carbon signature measured in this lake. The same explanation can be recalled for Lake Mergozzo, where a transition between *C. abyssorum* and the smaller-sized *M. leuckarti* and *Mesocyclops hyalinus* is known to occur in summer [40,41]. Δ^{15}N enrichment confirms some traits found in other lakes, independently from their size, typology, and trophy. In particular, cyclopoids were the most enriched in all but Lake Endine, in which the diaptomids were more enriched than the cyclopoids [20]. The small seasonal variability of δ^{15}N signature also suggests a small plasticity in trophic position in all lakes but Pusiano, the hypereutrophic lake. In this lake, the cyclopoids relied upon the pelagic signature also in summer.

5. Conclusions

Isotopic signatures can be regarded as fingerprints of lake ecosystems, integrating site-specific traits related to lake trophy and typology. According to our results, some traits, such as the trend toward less depleted ^{13}C and less enriched ^{15}N signatures in summer than in spring, were confirmed, although with some exceptions. While in thermally stratified lakes, such a transition was detected, in the shallowest, Lake Comabbio, the prevalence of a horizontal gradient was likely responsible for the opposite pattern. On the other hand, while the hypereutrophic Lake Pusiano was the most δ^{15}N enriched as expected, substantially high levels of δ^{15}N signature were also detected in the small, oligotrophic, and deep Lake Moro, the coldest lake among those analyzed.

Our results overall confirm the importance of taxa-specific analyses within zooplankton community, which is essential for understanding trophic relationships, changes in habitat, and carbon sources fueling the pelagic food web.

Supplementary Materials: The following are available online at http://www.mdpi.com/2073-4441/10/1/94/s1.

Author Contributions: All authors contributed equally to the paper.

Conflicts of Interest: The authors declare no conflict of interest. The founding sponsors had no role in the design of the study; in the collection, analyses, or interpretation of data; in the writing of the manuscript, and in the decision to publish the results.

References

1. Middelburg, J.J. Stable isotopes dissect aquatic food webs from the top to the bottom. *Biogeosciences* **2014**, *11*, 2357–2371. [CrossRef]
2. Peterson, B.J.; Fry, B. Stable isotopes in ecosystem studies. *Annu. Rev. Ecol. Syst.* **1987**, *18*, 293–320. [CrossRef]
3. Kling, G.W.; Fry, B.; O'Brien, W.J. Stable isotopes and planktonic trophic structure in arctic lakes. *Ecology* **1992**, *73*, 561–566. [CrossRef]
4. Cabana, G.; Rasmussen, J.B. Comparison of aquatic food chains using nitrogen isotopes. *Proc. Natl. Acad. Sci. USA* **1996**, *93*, 10844–10847. [CrossRef] [PubMed]
5. Post, D.M. Using stable isotopes to estimate trophic position: Models, methods, and assumptions. *Ecology* **2002**, *83*, 703–718. [CrossRef]
6. Visconti, A.; Manca, M. Seasonal changes in the δ^{13}C and δ^{15}N signatures of the Lago Maggiore pelagic food web. *J. Limnol.* **2011**, *70*, 263–271. [CrossRef]

7. Layman, C.A.; Araujo, M.S.; Boucek, R.; Hammerschlag-Peyer, C.M.; Harrison, E.; Jud, Z.R.; Matich, P.; Rosenblatt, A.E.; Vaudo, J.J.; Yeager, L.A.; et al. Applying stable isotopes to examine food-web structure: An overview of analytical tools. *Biol. Rev. Camb. Philos. Soc.* **2012**, *87*, 545–562. [CrossRef] [PubMed]

8. Fadda, A.; Manca, M.; Camin, F.; Ziller, L.; Buscarino, P.; Mariani, M.; Padedda, B.M.; Sechi, N.; Virdis, T.; Lugliè, A. Study on the suspended particulate matter of a Mediterranean artificial lake (Sos Canales Lake) using Stable Isotope Analysis of carbon and nitrogen. *Ann. Limnol. Int. J. Limnol.* **2016**, *52*, 401–412. [CrossRef]

9. Vander Zanden, M.J.; Casselman, J.M.; Rasmussen, J.B. Stable isotope evidence for the food web consequences of species invasions in lakes. *Nature* **1999**, *401*, 464–467. [CrossRef]

10. Grey, J.; Jones, R.I.; Sleep, D. Stable isotope analysis of the origins of zooplankton carbon in lakes of differing trophic state. *Oecologia* **2000**, *123*, 232–240. [CrossRef] [PubMed]

11. Cattaneo, A.; Manca, M.; Rasmussen, J.B. Peculiarities in the stable isotope composition of organisms from an alpine lake. *Aquat. Sci. Res. Bound.* **2004**, *66*, 440–445. [CrossRef]

12. Visconti, A.; Volta, P.; Fadda, A.; Di Guardo, A.; Manca, M. Seasonality, littoral versus pelagic carbon sources, and stepwise ^{15}N-enrichment of pelagic food web in a deep subalpine lake: The role of planktivorous fish. *Can. J. Fish. Aquat. Sci.* **2014**, *71*, 436–446. [CrossRef]

13. Post, D.M.; Pace, M.L.; Hairston, N.G. Ecosystem size determines food-chain length in lakes. *Nature* **2000**, *405*, 1047–1049. [CrossRef] [PubMed]

14. LIMNO Database Della Qualità dei Laghi Italiani. Available online: www.ise.cnr.it/limno/schede (accessed on 3 December 2004).

15. Uso del Suolo in Regione Lombardia-Atlante Descrittivo. Available online: http://www.ersaf.lombardia.it/upload/ersaf/gestionedocumentale/AtlanteUsoDelSuolo2010_784_15401.pdf (accessed on 10 May 2012).

16. Barbanti, L.; Bonacina, C.; Calderoni, A.; Carollo, A.; de Bernardi, R.; Guilizzoni, P.; Nocentini, A.M.; Ruggiu, D.; Saraceni, C.; Tonolli, L. *Indagini Ecologiche sul Lago d'Endine*; Edizioni dell'Istituto Italiano di Idrobiologia: Pallanza, Italy, 1974; 304p.

17. Garibaldi, L.; Brizzio, M.C.; Varallo, A.; Mosello, R. The improving trophic conditions of Lake Endine (Northern Italy). *Mem. Ist. Ital. Idrobiol.* **1997**, *56*, 23–36.

18. Garibaldi, L.; Mosello, R.; Brizzio, M.C.; Varallo, A. *Chimica e Fitoplancton del Lago Moro (Alpi Orobiche Bresciane)*; Picazzo, M., Ed.; Atti A.I.O.L.: Genova, Italy, 2000; Volume 13, pp. 395–410.

19. Osservatorio dei Laghi Lombardi. Qualità delle acque lacustri in Lombardia. In *Primo Rapporto OLL 2004*; Osservatorio dei Laghi Lombardi: Brugherio, Italy, 2005; 351p.

20. Fadda, A.; Rawcliffe, R.; Padedda, B.M.; Lugliè, A.; Sechi, N.; Camin, F.; Ziller, L.; Manca, M. Spatiotemporal dynamics of C and N isotopic signature of zooplankton: A seasonal study on a man-made lake in the Mediterranean region. *Ann. Limnol. Int. J. Limnol.* **2014**, *50*, 279–287. [CrossRef]

21. Caroni, R.; Free, G.; Visconti, A.; Manca, M. Phytoplankton functional traits and seston stable isotopes ratio: A functional-based approach in a deep, subalpine lake, Lake Maggiore (N. Italy). *J. Limnol.* **2012**, *71*, 84–94. [CrossRef]

22. Cabana, G.; Rasmussen, J.B. Modeling food chain structure and contaminant bioaccumulation using stable nitrogen isotopes. *Nature* **1994**, *372*, 255–257. [CrossRef]

23. Degens, E.T.; Guillard, R.; Sackett, W.M.; Hellebust, J.A. Metabolic fractionation of carbon isotopes in marine plankton. Temperature and respiration experiments. *Deep Sea Res.* **1968**, *15*, 1–9. [CrossRef]

24. Fry, B.; Wainright, S.C. Diatom sources of ^{13}C-rich carbon in marine food webs. *Mar. Ecol. Prog. Ser.* **1991**, *76*, 149–157. [CrossRef]

25. Zohary, T.; Erez, J.; Gophen, M.; Berman-Frank, I.; Stiller, M. Seasonality of stable carbon isotopes within the pelagic food web of Lake Kinneret. *Limnol. Oceanogr.* **1994**, *39*, 1030–1104. [CrossRef]

26. France, R.L.; Del Giorgio, P.A.; Westcott, K.A. Productivity and heterotrophy influences on zooplankton ^{13}C in northern temperate lakes. *Aquat. Microb. Ecol.* **1997**, *12*, 85–93. [CrossRef]

27. Leggett, M.F.; Servos, M.R.; Hesslein, R.; Johannsson, O.; Millard, E.S.; Dixon, D.G. Biogeochemical influences on the carbon isotope signatures of Lake Ontario biota. *Can. J. Fish. Aquat. Sci.* **1999**, *56*, 2211–2218. [CrossRef]

28. Leggett, M.F.; Johannsson, O.; Hesslein, R.; Dixon, D.G.; Taylor, W.D.; Servos, M.R. Influence of inorganic nitrogen cycling on the δ^{15}N of Lake Ontario biota. *Can. J. Fish. Aquat. Sci.* **2000**, *57*, 1489–1496. [CrossRef]

29. Ruggiu, D.; Morabito, G.; Panzani, P.; Pugnetti, A. Trends and relations among basic phytoplankton characteristics in the course of the long-term oligotrophication of Lake Maggiore (Italy). *Hydrobiologia* **1998**, *369*, 243–257. [CrossRef]

30. Matthews, B.; Mazumder, A. Compositional and interlake variability of zooplankton affect baseline stable isotope signature. *Limnol. Oceanogr.* **2003**, *48*, 1977–1987. [CrossRef]

31. Marty, J.; Planas, D. Comparison of methods to determine algal δ^{13}C in freshwater. *Limnol. Oceanogr. Meth.* **2008**, *6*, 51–63. [CrossRef]

32. Perga, M.E.; Gerdeaux, D. Changes in the δ^{13}C of pelagic food webs: The influence of lake area and trophic status on the isotopic signature of whitefish (*Coregonus lavaretus*). *Can. J. Fish. Aquat. Sci.* **2004**, *61*, 1485–1492. [CrossRef]

33. Hamza, W.; Ruggiu, D.; Manca, M. Diel zooplankton migrations and their effect on the grazing impact in Lake Candia (Italy). *Arch. Hydrobiol.* **1993**, *39*, 175–185.

34. Syväranta, J.; Haemaelaeinen, H.; Jones, R.I. Within-lake variability in carbon and nitrogen stable isotope signatures. *Freshwat. Biol.* **2006**, *51*, 1090–1102. [CrossRef]

35. Grey, J.; Jones, R.I.; Sleep, D. Seasonal changes in the importance of the source of organic matter to the diet of zooplankton in Loch Ness, as indicated by stable isotope analysis. *Limnol. Oceanogr.* **2001**, *46*, 505–513. [CrossRef]

36. Minagawa, M.; Wada, E. Stepwise enrichment of ^{15}N along food chains: Further evidence and the relation between δ^{15}N and animal age. *Geochim. Cosmochim.* **1984**, *48*, 1135–1140. [CrossRef]

37. Santer, B.; Sommerwerk, N.; Grey, J. Food niches of cyclopoid copepods in eutrophic Plußsee determined by stable isotope analysis. *Arch. Hydrobiol.* **2006**, *167*, 301–316. [CrossRef]

38. Van den Bosch, F.; Santer, B. Cannibalism in *Cyclops abyssorum*. *Oikos* **1993**, *67*, 19–28. [CrossRef]

39. Santer, B.; Van den Bosch, F. Herbivorous nutrition of *Cyclops vicinus*: The effect of a pure algal diet on feeding, development, reproduction and life cycle. *J. Plankton Res.* **1994**, *16*, 171–195. [CrossRef]

40. De Bernardi, R.; Soldavini, E. Long-term fluctuations of zooplankton in Lake Mergozzo, Northern Italy. *Mem. Ist. Ital. Idrobiol.* **1976**, *33*, 345–375.

41. De Bernardi, R.; Soldavini, E. Seasonal dynamics of the zooplankton community in Lago di Mergozzo (Northern Italy). *Mem. Ist. Ital. Idrobiol.* **1977**, *34*, 137–154.

Article

Sedimentary Record of Cladoceran Functionality under Eutrophication and Re-Oligotrophication in Lake Maggiore, Northern Italy

Liisa Nevalainen [1,*], Meghan Brown [2] and Marina Manca [3]

[1] Ecosystems and Environment Research Programme, Faculty of Biological and Environmental Sciences, University of Helsinki, Niemenkatu 73, 15140 Lahti, Finland

[2] Department of Biology, Hobart and William Smith Colleges, 4095 Scandling Center, Geneva, NY 14456, USA; mbrown@hws.edu

[3] National Research Council, Water Research Institute, Largo Tonolli 50, 28922 Verbania, Italy; m.manca@ise.cnr.it

[*] Correspondence: liisa.nevalainen@helsinki.fi; Tel.: +358-2941-20311

Received: 20 December 2017; Accepted: 16 January 2018; Published: 19 January 2018

Abstract: We examined fossil Cladocera (Crustacea) communities and their functional assemblages in a ~60-year sediment record from Lake Maggiore, northern Italy. Our main objective was to document the response of aquatic community functioning to environmental stress during eutrophication (1960–1985) and recovery (post-1985), and to identify environmental controls on cladoceran functionality. Of the functional groups, large filter feeders and oval epibenthos thrived prior to eutrophication (reference conditions pre-1960) and globular epibenthos and small filter feeders increased during eutrophication and as the lake recovered. Multivariate analyses suggested that bottom-up controls (i.e., total phosphorus) were important for shaping functional assemblages but taxonomic community changes were likely related to top-down control by predators, particularly the predaceous cladoceran *Bythotrephes longimanus*. Functional diversity (FD) was higher and *Daphnia* ephippia length (DEL) larger during the reference and early eutrophication periods and decreased during eutrophication and recovery. Both FD (high) and DEL (large) were distinct during reference period, but were similar (FD low, DEL small) between the eutrophication and recovery periods. The functional attributes and the assemblages did not recover post-eutrophication, suggesting that the system exhibited a clear shift to low FD and dominance of small filterers. Cladoceran functionality appears to be related to fundamental ecosystem functions, such as productivity, and may thus provide insights for long-term changes in ecological resilience.

Keywords: biodiversity; environmental change; fossil Cladocera; functional diversity; paleolimnology; ecological resilience; subalpine lakes

1. Introduction

Nutrient enrichment of freshwaters is a worldwide challenge [1]. In combination with intensified climate warming, these anthropogenic changes threaten aquatic biodiversity and ecosystem services [2,3]. Ecological communities and their responses to environmental stressors, such as eutrophication in aquatic systems, can be investigated with a functional approach, where interest is put on species' ecological roles. For example, feeding traits, habitat preferences, reproduction, or morphological attributes (e.g., body size) can reflect certain ecological functions [4]. The concept of functionality in ecological communities allows a comprehensive understanding of how environmental changes alter ecosystems through biological functions rather than just taxonomic composition [5]. This kind of mechanistic approach on aquatic systems may reflect important ecosystem level processes, for example, changes in productivity and trophic structure of lakes. In the functional approach, functional diversity (FD) is a biodiversity measure, which takes into account the variety of biological functions of species and may allow for a more holistic understanding of environmental changes and ecosystem responses [4,6].

With major losses in global biodiversity [7], paleolimnological data sets can aid our understanding of the relationships among functional diversity and ecosystem productivity, climate change, and trophic dynamics [8–10]. Since key members of aquatic communities (or their traces) are preserved as fossils in lake sediments, paleolimnology can be used to evaluate ecosystem functions and ecological resilience in lakes. Further, lake sediment archives are advantageous for functional classification and ecosystem level responses to environmental changes, because time lags often prevent detection during short-term observations [11,12].

Here, we continue the application of paleolimnological research on Lake Maggiore in northern Italy (Figure 1) to investigate the lake as a "natural laboratory" with its well-documented history of eutrophication and re-oligotrophication during the last century. This subalpine lake is Italy's second largest and deepest lake and part of the long-term limnological monitoring in Europe [13]. Naturally oligotrophic and phosphorus-limited Lake Maggiore eutrophied in the 1960s as a result of nutrient loading from the catchment and wastewater discharge [14,15]. Phosphorus concentrations started to increase and the peak of nutrient enrichment occurred during the late 1970s when total phosphorus at winter mixing (TP_{mix}) was 31 µg L^{-1} and the lake became mesotrophic (Figure 2). After that, due to enhancements in wastewater treatment, recovery of the lake proceeded and TP decreased back to oligotrophic levels (~10 µg L^{-1}) during the early 1990s. Previous research on Lake Maggiore plankton has indicated that the aquatic communities are highly responsive to nutrient status, trophic dynamics, and climate [15–18]. For example, Lake Maggiore zooplankton has exhibited major changes in taxonomic composition, body size, and population density under eutrophication and re-oligotrophication [15,17,18].

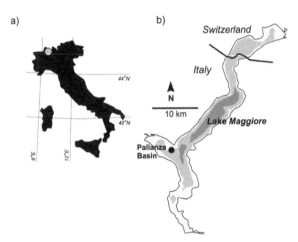

Figure 1. (a) Location of Lake Maggiore in northern Italy (gray dot) and (b) location of the sediment core sampling site in the Pallanza Basin of Lake Maggiore (black dot), where the shades of gray represent bathymetric details (light gray < 100 m, mid gray 100–300 m, and dark gray > 300 m water depth).

The aim of the current research was to examine responses to eutrophication and the subsequent limnological recovery of Lake Maggiore by the cladoceran communities. We analyzed fossil cladoceran communities for their taxonomic composition, functional characterization, and functional attributes of FD and *Daphnia* ephippia length (DEL) in a sediment core covering the years of pre-eutrophication, eutrophication and recovery (1944–2010). We aimed to identify the main environmental forcings on the long-term succession of cladoceran communities, functional assemblages, and functional attributes, and discuss the roles of bottom-up versus top-down controls in ecosystem functioning during the eutrophication and re-oligotrophication of Lake Maggiore.

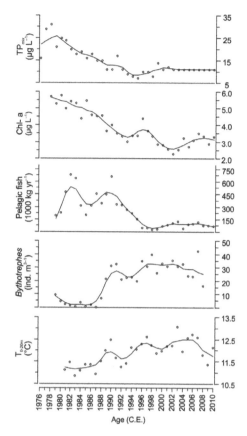

Figure 2. Real time observations in the long-term changes in Lake Maggiore total phosphorus at winter mixing (TP$_{mix}$, since 1976), chlorophyll-a concentration (Chl-a, since 1978), pelagic fish catch (since 1979), *Bythotrephes longimanus* abundances (1979–2008) and water temperature of the euphotic zone 0–20 m (T$_{0-20m}$, since 1981). These data are previously partly published [15,18].

2. Materials and Methods

A modified Wilco box corer (liner internal surface area 28 cm^2, shaft depth 50.5 cm) was used to collect a 34-cm core in the Pallanza Basin of Lake Maggiore (45°54.76 N, 8°32.96 E, z = 98 m) on 11 February 2010 (Figure 1). The intact core was stored dark at 4 °C until processing. The core was vertically sliced at 1-cm intervals and the outer edge of each segment was discarded. A 10 g subsample was removed from each interval for ^{137}Cs dating at University of Applied Sciences (Holland). Increases in the ^{137}Cs activity were interpreted as increased fallout of ^{137}Cs from nuclear activity (Figure 3a). An age–depth model was created by interpolating the surface sediment sample (0 cm, year 2010), increased ^{137}Cs activities of Chernobyl fallout (1986) at 14 cm, and the nuclear weapons testing horizon (1963) at 25 cm (Figure 3b). As the model is based on constant sedimentation rate between these time horizons it should be interpreted cautiously.

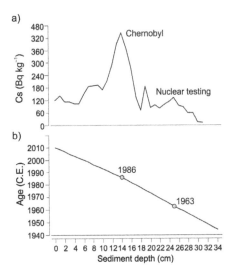

Figure 3. (**a**) ^{137}Cs activity and (**b**) an age-depth model for the Lake Maggiore sediment core. Peak activities of Chernobyl nuclear disaster (year 1986) and nuclear weapon testing (1963) are indicated.

The remaining portion of each sediment interval was prepared for fossil Cladocera analysis following the standard methods [19]. First, the samples were heated in 10% KOH to deflocculate sediment and then sieved through a 76-µm mesh, which is adequate to retain cladoceran zooplankton fossil remains and the remains of smallest taxa of Chydoridae (e.g., *Alonella nana* and *Coronatella rectangula*, Table 1). The residue was stained pink with safranine. The fossil Cladocera remains (carapaces, headshields, postabdomens and ephippia) were identified and enumerated with a light microscope (magnifications 100–400X) and the most abundant body part was chosen to represent the number of individuals of each species. A minimum of 100 fossil individuals were counted from each subsample. Relative abundances of individual taxa were used to determine the community composition in the core samples. In addition, to examine functional characterization of the community (i.e., functional assemblages) taxa were assigned to the groups (Table 1): predators, large filter feeders, small and intermediate filter feeders (hereafter small filterers), globular epibenthos, and oval epibenthos based on a previous functional grouping of Cladocera [10]. Size (i.e., body length) from base to apex (spine excluded) of encountered fossil *Daphnia* ephippia was measured with a Zeiss microscope at 100X magnification equipped with a camera and analyzed with Image pro express 5 software to estimate the mean *Daphnia* ephippia length (DEL). Ephippia were measured from 32 sediment samples and number of size measurements per sediment subsection varied from 1 to 73 (mean 13).

Principal component analysis (PCA) was used to summarize temporal succession of cladoceran taxonomic communities (compositional gradient <1.5 SD, standard deviation units) and functional assemblages (<1.0 SD). The response data were square root transformed for PCA. In addition, redundancy analysis (RDA) was used to analyze relationships between functional assemblages and cladoceran taxonomic communities and limnological variables during the period of continuous environmental monitoring (since 1978). Chlorophyll-*a* concentration (chl-*a*), total phosphorus at winter mixing (TP_{mix}), water temperature of the euphotic zone 0–20 m (T_{0-20m}), *Bythotrephes longimanus* abundance, and total pelagic fish catch were included as environmental variables (Figure 2). Environmental variables were forward selected and the significance ($p \leq 0.05$) of each variable was tested with Monte Carlo permutations (999). Cladoceran functional diversity (FD) was evaluated with Rao's FD index [20], i.e., Rao's quadratic entropy. For the index, each cladoceran taxa was assigned with qualitative functional character including body size (small < 500 µm, intermediate 500–1000 µm, large > 1000 µm), body shape (elongated, oval, globular), feeding type (filterer, scraper-detritivore, predator) and microhabitat (pelagic, benthic, attached to vegetation, Table 1) [10]. This characterization was based on ecological data available for cladoceran taxa [21,22] and the characters were inserted as functional character present (1) and absent (0). Multivariate analyses (PCA and RDA) and analysis of FD were performed with Canoco 5 software [23]. Tukey's pairwise comparisons were utilized to indicate differences in FD and DEL during the reference (pre-1960), eutrophication (1960–1985), and recovery (post-1985) periods. These analyses were performed with PAST software [24]. Segmented regression analysis was utilized to detect statistically significant breakpoints (minimum confidence level of 95%) in FD and DEL. The best breakpoint was selected based on maximizing the statistical coefficient of explanation and performing tests of significance with SegReg program [25].

Table 1. Functional characterization of cladoceran taxa (indicated with asterisks) encountered from Lake Maggiore sediment core based on body size (S = small, M = intermediate, L = large), body shape (G = globular, O = oval, E = elongated), feeding type (F = filterer, S-D = scraper-detritivore, P = predator, including parasitism), and habitat (P = pelagial, B = benthic, V = vegetation) and their functional grouping (FG) used in Figure 4. Among each functional group, the taxa are listed according to their mean relative abundance from the most abundant at the top.

	Body Size			Body Shape			Feeding Type			Habitat			
	S	M	L	G	O	E	F	S-D	P	P	B	V	FG
Leptodora kindtii			*			*			*	*			Predator
Sida crystallina		*			*		*			*		*	Large filterer
Daphnia longispina-type		*			*		*			*			Large filterer
Eubosmina longispina-type	*			*			*			*			Small filterer
Eubosmina coregoni-type	*			*			*			*			Small filterer
Bosmina longirostris	*			*			*			*			Small filterer
Chydorus cf. *sphaericus*	*			*				*		*		*	Globular epibenthos
Paralona pigra	*			*				*			*		Globular epibenthos
Pleuroxus uncinatus		*		*				*			*	*	Globular epibenthos
Monospilus dispar	*			*				*			*		Globular epibenthos
Alonella excisa	*			*				*			*	*	Globular epibenthos
Pleuroxus trigonellus		*		*				*			*	*	Globular epibenthos
Alonella exigua	*			*				*				*	Globular epibenthos
Alonella nana	*			*				*			*	*	Globular epibenthos
Anchistropus emarginatus	*			*					*		*	*	Globular epibenthos
Alona affinis	*				*			*			*	*	Oval epibenthos
Alona quadrangularis	*				*			*			*		Oval epibenthos
Acroperus harpae	*				*			*			*	*	Oval epibenthos
Eurycercus spp.		*			*			*			*	*	Oval epibenthos
Alona guttata	*				*			*			*	*	Oval epibenthos
Camptocercus rectirostris		*			*			*			*	*	Oval epibenthos
Disparalona rostrata		*			*			*			*		Oval epibenthos
Peracantha truncata	*				*			*			*	*	Oval epibenthos
Coronatella rectangula	*				*			*		*	*	*	Oval epibenthos
Alona intermedia	*				*			*			*	*	Oval epibenthos
Graptoleberis testudinaria		*			*			*				*	Oval epibenthos
Pleuroxus laevis		*			*			*			*	*	Oval epibenthos
Alona rustica		*			*			*			*	*	Oval epibenthos
Alona guttata f. *tuberculata*	*				*			*			*	*	Oval epibenthos
Rhyncotalona falcata	*				*			*			*		Oval epibenthos

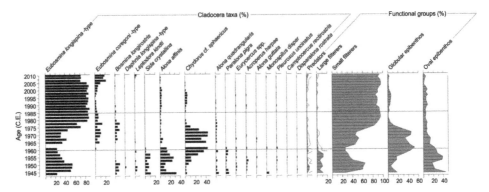

Figure 4. Relative abundances of most abundant fossil cladoceran taxa and functional groups in Lake Maggiore sediment core. The horizontal lines define the reference (pre-1960), eutrophication (1960–1985), and recovery (post-1985) periods.

3. Results

We detected 30 cladoceran taxa in the Lake Maggiore sediment core (Table 1). The most abundant taxa were Bosminidae, including *Eubosmina longispina*-type (*n* = 34, mean percent abundance 63.5%), *Eubosmina coregoni*-type (*n* = 26, 5.7%), and *Bosmina longirostris* (28, 2.6%) and Chydoridae (chydorids), including *Chydorus* cf. *sphaericus* (*n* = 31, 11.3%) and *Alona affinis* (*n* = 34, 6.7%). In the early core (until 1960s; reference period) the communities were dominated by *A. affinis* and *E. longispina*-type (~60%) together with lower abundances of *B. longirostris*, *Leptodora kindtii*, *Sida crystallina* and several less common chydorids (e.g., *Alona quadrangularis*, *Paralona pigra*, *Eyrycercus* spp. and *Acroperus harpae*; Figure 4). *Chydorus* cf. *sphaericus* started to increase in the early 1960s and reached maximum abundance (~40%) and dominance during early 1970s. *Bythotrephes longimanus*, which were abundant in zooplankton samples starting in the 1990s (Figure 2), were not detected in the fossil community (Table 1). In functional assemblages, succession proceeded from oval epibenthos during pre-1960s to globular epibenthos (~40–50%) between 1960 and 1975, and small filterers (80–90%) after that until the top core (Figure 4). Predators and large filterers were scarce (<10%) and occurred mostly prior to 1960s.

PCA for cladoceran taxonomic communities resulted in eigenvalues 0.514 for PCA axis 1 and 0.217 for PCA axis 2. Cumulative percentage of variance explained by the PCA were 51.4% and 73.2% for PCA axes 1 and 2, respectively. Samples in the reference period (pre-1960) had positive PCA axis 1 scores with increasing PCA axis 2 scores (Figure 5a). Samples from the eutrophication period (1960–1985) had reducing scores along PCA axes 1 and 2. Recovery period samples (post-1985) clumped together at the negative end of PCA axis 1. PCA for functional assemblages had eigenvalues of 0.763 for PCA axis 1 (cumulative % of variance 76.3) and 0.169 for axis 2 (93.2%). Sample scores drifted from positive axis 2 values to negative prior to 1960 and from positive to negative axis 1 values during eutrophication (Figure 5b). Most recent samples of the recovery period had negative axis 1 values and close to zero axis 2 values.

RDA for functional assemblages resulted in eigenvalues 0.222 for RDA axis 1 and 0.1585 for axis 2 and all the environmental variables explained 40.6% of the variance in the assemblages. RDA identified TP_{mix} (38.3%) as the single significant ($p < 0.05$) environmental factors explaining variance in functional assemblages (Table 2). RDA for taxonomic communities resulted in eigenvalues 0.2158 and 0.0778 for axes 1 and 2, respectively and forward selection did not results in any statistically significant results (Table 2). *Bythotrephes longimanus* abundances, however, explained most variance in the communities with 33.4% ($p = 0.0600$).

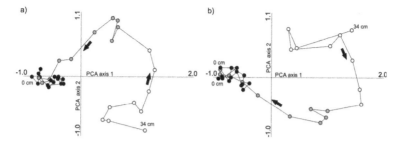

Figure 5. Principal component analysis (PCA) ordination diagrams for (**a**) cladoceran communities and (**b**) functional assemblages, where white symbols = reference period (pre-1960), gray symbols = eutrophication period (1960–1985), and black symbols = recovery (post-1985). The top (0 cm) and bottom (34 cm) samples of the core are marked, the black lines connect sample points chronologically, and the arrows indicate the direction of change.

Table 2. Forward selection statistic (% explained, F- and *p*-values) of environmental variables (see abbreviations in Figure 2) in redundancy analysis (RDA) applied separately for cladoceran functional assemblages and taxonomic communities (and percentage of variance explained by all environmental variables) in Lake Maggiore since the period of continuous environmental monitoring. Most significant environmental variables are marked in bold type.

	Functional Assemblages (40.6% Explained by All Var.)	**Taxonomic Communities** (38.7% Explained by All Var.)
TP_{mix}	**38.3% ($F = 2.58$, $p = 0.039$)**	13.2% ($F = 0.85$, $p = 0.515$)
Chl-*a*	23.7% ($F = 1.67$, $p = 0.183$)	12.2% ($F = 0.77$, $p = 0.569$)
Bythotrephes	19.6% ($F = 1.43$, $p = 0.247$)	**33.4% ($F = 2.13$, $p = 0.060$)**
T_{0-20m}	16.0% ($F = 1.19$, $p = 0.327$)	20.3% ($F = 1.33$, $p = 0.220$)
Fish	2.35% ($F = 0.16$, $p = 0.954$)	20.9% ($F = 1.23$, $p = 0.242$)

FD varied between 1.7 and 3.2 in Rao's FD index (Figure 6). The highest FD occurred in the early core until 1960 after which FD decreased until the early 1980s. FD increased slightly between 1990 and 2000 but was reduced post-2000. DEL measurements varied between 470 and 690 μm (Figure 6). DEL was largest (~650 μm) in the early to mid-core until 1970s and then started to decrease and remained consistent (~550 μm). Tukey's test indicated significant differences in FD and DEL between reference and eutrophication periods, and reference and recovery periods (Table 3). No significant differences in FD or DEL were found between eutrophication and recovery periods. SegReg identified single significant breakpoints for FD at 15.64 cm (early 1980s) and DEL at 20.14 (early 1970s, Figure 6).

Table 3. Tukey's pairwise comparisons (Q-value, *p*-values in brackets) of cladoceran functional diversity (FD) and mean *Daphnia* ephippia length (DEL) between periods of recovery (pre-1960), eutrophication (1960–1985), and recovery (post-1985). Statistically significant differences are marked in bold type.

	Recovery	**Eutrophication**	**Reference**
Recovery	-	[DEL] 1.965 (0.360)	**[DEL] 7.828 (<0.001)**
Eutrophication	[FD] 3.371 (0.059)	-	**[DEL] 5.863 (<0.001)**
Reference	**[FD] 11.23 (<0.001)**	**[FD] 7.863 (<0.001)**	-

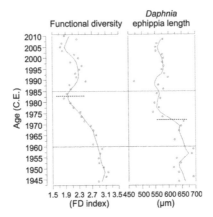

Figure 6. Cladoceran functional diversity (FD) and mean *Daphnia* ephippia length in the Lake Maggiore sediment core. The horizontal lines define the reference (pre-1960), eutrophication (1960–1985), and recovery (post-1985) periods, and the dashed horizontal lines represent segmented regression breakpoints.

4. Discussion

Bottom-up (food, habitats) and top-down (predators) controls drive cladoceran communities in Lake Maggiore, based on trophic dynamics inferred from sediments dating from 1943 to 2002 [17]. Here we present more refined functional analysis and include the most recent decade of fossil records (1944–2010). Cladoceran communities succeeded from *Alona affinis* to *Chydorus* cf. *sphaericus*, and later to *Eubosmina longispina*-type dominance with a functional shift from large filterers (e.g., *Sida*, *Daphnia*) and oval epibenthos (e.g., *A. affinis*) to globular epibenthos (*C.* cf. *sphaericus*) and most recently to small filterers (*Eubosmina*, Figure 4). The biotic changes occurred gradually and even prior to the eutrophication as small filterers and epibenthos exhibited shifts prior to 1960s but predators and large filterers responded promptly at ~1960 (Figure 4). This suggests that there was a difference in timing of the response among the functional groups and species, likely related to changed resources just prior to major nutrient loading. The PCAs indicated a clear temporal shift in the taxonomic communities and functional assemblages through the reference–eutrophication–recovery phases with a new stable state during the recent years of re-oligotrophication (Figure 5). The opposite drift of PCA axis 2 scores in taxonomic (from negative to positive scores, Figure 5a) and functional assemblages (from positive to negative, Figure 5b) during the reference–early eutrophication may suggest that diverse control mechanisms drove changes in taxonomic vs. functional communities. In agreement, RDA forward selection statistics suggest that functional assemblage shifts were associated with bottom-up controls (Table 2), since they were mainly predicted by TP$_{mix}$ during the time period of the core covered by limnological monitoring (since late 1970s, Figure 2). Taxonomic communities were best explained, although only marginally significant, with top-down control, i.e., *Bythotrephes* abundance (Table 2). *Bythotrephes* is known to regulate zooplankton, especially *Daphnia* as its main prey item, in Lake Maggiore and impact their distribution, abundance and phenology [18,26,27].

In general, caution should be taken when applying the ecological patterns reported here as whole-lake patterns, as the current sediment core was sampled from the Pallanza Basin (Figure 1). Nevertheless, the trends reported here are consistent with the previous research from another core [17]. Fossil remains (e.g., mandibles, caudal spines, resting eggs) of *Bythotrephes longimanus*, although abundant in zooplankton samples in Lake Maggiore (Figure 2, also in the Pallaza Basin), were not detected from the current core (Table 1) or in the previous sediment-based research [17] and the reason for that remains unexplained. Typically, *Bythotrephes* remains (caudal spines and sometimes resting eggs) are well recovered from sediments and their accumulation has been used to estimate presence and abundance [28–31]. *Bythotrephes* remains have been found from sediments in large and deep subalpine lakes, such as Lake Garda in northern Italy [32], in similar type of geo-limnological settings than Lake Maggiore. The use of a larger (76-µm) than the commonly used ~50-µm sieve [19] for processing sediment samples for fossil cladoceran analysis in the current study is not a reasonable explanation for the lack of *Bythotrephes* remains, since the previous studies have used even a 250-µm aperture sieve [28–30]. Accordingly, it may be that the sediment sampling site in the Pallanza Basin was not a representative location for *Bythotrephes* fossils to preserve. It is further possible that some still unidentified limnological factor in Lake Maggiore prevents *Bythotrephes* fossils for preserving in the sediments.

Functional assemblages of the Lake Maggiore core succeeded from epibenthos to small filterer dominance during the eutrophication and re-oligotrophication (Figure 4). Globular epibenthos increased in conjunction with eutrophication succession around 1970s. This group is mainly formed by a single dominant taxon (*C.* cf. *sphaericus*), whose feeding is characterized by fine-mesh filtering [33]. It is adapted to diverse resources of detrital food (detritus and attached microbes) and pelagic habitats under eutrophic conditions [34,35] and has been reported to occur at high abundances during the eutrophication period of Lake Maggiore [17]. A stable functional phytoplankton community dominated by large-celled groups was present during the late eutrophication and early recovery [15]. Clear responses of the phytoplankton communities to reduced nutrients started to occur during late 1980s when small-sized phytoplankton groups increased in the lake [15]. The functional succession in Lake Maggiore zooplankton toward small planktonic filterers (i.e., *Eubosmina*) may have been partly related to the more efficient grazing of these small sized planktonic cladocerans under low-density food conditions [36,37], i.e., oligotrophic conditions. However, relying on the current chronology (Figure 3), functional assemblages of Cladocera responded earlier than the phytoplankton (late 1980s) to nutrient reductions as small filterers increased already during the early 1980s (Figure 4) [15]. Therefore, it is likely that the shifts in functional assemblages in Lake Maggiore were also affected by varying vulnerability of different functional groups to predation and were partly driven by top-down controls, although not identified by the RDA (Table 2). Pelagic fish abundances increased in the early 1980s (Figure 2) and fish tend to prey on largest and most visible prey [38] fitting with the increase of small filterers in Lake Maggiore.

We detected a significant breakpoint in FD during the early 1980s (segmented regression breakpoint, Figure 6), which occurred in conjunction with the lowering trend in TP_{mix} at the turn of the decade (Figure 2). FD started to decline during the 1960s under eutrophication, whereas it was highest during the pre-eutrophication period suggesting a negative relationship between lake productivity and FD (Figure 6). Contrasting results were found from small and shallow boreal lakes in Finland [10], where FD of cladocerans experienced a long-term positive relationship with eutrophication until mesotrophic conditions (TP ~40–60 µg L^{-1}). The cause for such a positive relationship was apparently diversifying niche space in both pelagic and littoral habitats until mesotrophic conditions, but after which deterioration of littoral-benthic resources reduced FD [10]. A different niche space pattern (i.e., high niche space during low productivity and vice versa) may explain the negative FD–productivity relationship in Lake Maggiore during the eutrophication. This variable pattern is likely related to divergent habitat diversity dynamics in large and deep subalpine vs. small and shallow boreal lakes. For example, the littoral zones, with highest and most diverse niche space, of deep and steep basin morphology and transparent oligotrophic waters of Lake Maggiore may reduce rapidly by even slight changes in primary production and consequent changes in the euphotic zone depth. There exists a positive zooplankton FD–ecosystem function relationship in experimental systems [39] and, despite the geographical differences, the paleolimnological evidence from natural lake ecosystems and cladoceran fossils support the FD–ecosystem function (e.g., productivity) relationship.

The current DEL data series of Lake Maggiore can be considered to represent an estimate of a mean *Daphnia* ephippium size in the past. It should be interpreted with caution because ephippia were absent or scarce in some sediment samples because *Daphnia* was mostly parthenogenetic during the mesotrophic status in Lake Maggiore (M. Manca, unpublished data). DEL was larger during the pre-eutrophication conditions and reduced under eutrophication and was separated by a clear threshold in the early 1970s (segmented regression breakpoint, Figure 6). The limnological monitoring does not extend continuously back to 1970s but $TP_{mix} > 30\ \mu g\ L^{-1}$ were recorded during the late 1970s (Figure 2). Previously, it has been suggested that *Daphnia* body size decreases during eutrophication and increased abundance of planktivorous fish [40] fitting well with the Maggiore DEL data. From the bottom-up point of view, DEL development and its reduction was likely related to changes in food quality for *Daphnia* under oligotrophic conditions and competition with Bosminidae that are better adapted to low food levels and small-sized phytoplankton [36,37]. However, the breakpoint in DEL does not match with the planktonic rearrangement that took place in Maggiore during the 1980s [15] but precedes it with almost a decade suggesting that DEL was likely under top-down control.

Paleolimnological records on FD are valuable, but scarce, to understand ecosystem function–environmental stress interconnections and tipping points [10,12]. In connection with ecosystem function, the use of FD indices can indicate ecosystem resilience under environmental perturbations, such as eutrophication or climate warming. Since our Lake Maggiore core did not reach actual pre-disturbance conditions before any significant human impact, it is difficult to assess the natural level and fluctuation of FD as fish introductions started already in the 19th century. There was a regime shift visible in functional assemblages and attributes in the current record around 1980, suggesting a shift in ecosystem status that was evident also in the environmental data with reductions in primary production (Figure 2) and planktonic algal communities [15]. This was evident for cladoceran communities described in previous investigations, where a completely new community composition prevailed after eutrophication [17]. The functional assemblages have yet to return to pre-eutrophication conditions and the functional changes appear to be irreversible. In combination, FD and DEL did not return to pre-eutrophication high values and did not differ significantly during eutrophication and recovery phases (Figure 6, Table 3) suggesting that, despite reduced nutrient status, diversity and ecosystem functioning exhibited a new state.

The observed changes in functional attributes (Figures 4 and 6), even though principally explained by productivity in Lake Maggiore, may also include a signal of climate-induced shifts. Climate change may be difficult to decouple from those of eutrophication and re-oligotrophication in Lake Maggiore since some previously reported functional changes, e.g., dominance of smaller phytoplankton can also be explained by warming lake water in addition to re-oligotrophication [15], and these two environmental changes overlap temporally. Previous results from high alpine lakes have shown that limnological consequences of climate warming, e.g., deepening of thermocline and increasing water temperature may induce functional changes in cladocerans and reduce their FD [12]. Accordingly, thermal changes in the water column of Lake Maggiore, e.g., increasing epilimnetic water temperatures (Figure 2) or deeper thermocline, may be, at least in part, responsible for the functional changes, e.g., increase and dominance of small filter feeders and low FD.

In conclusion, FD and DEL had negative relationships with nutrient status in Lake Maggiore during the eutrophication and re-oligotrophication. A threshold was reached around 1980–1985 when cladoceran functionality changed resulting in higher abundance of small-sized planktonic grazers and lower FD and decreased DEL. Long-term development of FD was controlled mainly by bottom-up drivers and is therefore related to ecosystem functioning, i.e., productivity. More spatial and long-term temporal paleolimnological data is required to understand the phenomenon comprehensively. For example, understanding whether the observed pattern in FD and in-lake production is common only during anthropogenic eutrophication process and in large subalpine lakes or if such relationship exists among a geographically wider set of lakes and under natural lake ontogenic development.

Acknowledgments: Funding for this project included the VIOLET and SCUM projects (287547 and 308954) of the Academy of Finland to L. Nevalainen and a Fulbright research award to M. Brown. Interpretation of the sediment chronology was provided by Piero Guilizzoni and Victoria Putyrskaya. We acknowledge the constructive comments by Tomi Luoto and three anonymous reviewers.

Author Contributions: Meghan Brown and Marina Manca sampled the sediments and pretreated the samples; Liisa Nevalainen, Meghan Brown and Marina Manca analyzed the data; Liisa Nevalainen wrote the paper with comments from Meghan Brown and Marina Manca.

Conflicts of Interest: The authors declare no conflict of interest.

References

1. Schindler, D.W. The dilemma of controlling cultural eutrophication of lakes. *Proc. R. Soc. B* **2012**, *279*, 4322–4333. [CrossRef] [PubMed]

2. Heino, J.; Virkkala, R.; Toivonen, H. Climate change and freshwater biodiversity: Detected patterns, future trends and adaptations in northern regions. *Biol. Rev.* **2009**, *84*, 39–54. [CrossRef] [PubMed]

3. Whitehead, P.G.; Wilby, R.L.; Battarbee, R.W.; Kernan, M.; Wade, A.J. A review of the potential impacts of climate change on surface water quality. *Hydrol. Sci. J.* **2009**, *54*, 101–123. [CrossRef]

4. Hooper, D.U.; Solan, M.; Symstad, A.; Díaz, S.; Gessner, M.O.; Buchmann, N.; Degrande, V.; Grime, P.; Hulot, F.; Mermillod-Blondin, F.; et al. Species diversity, functional diversity, and ecosystem functioning. In *Biodiversity and Ecosystem Functioning. Synthesis and Perspectives*; Loreau, M., Naeem, S., Inchausti, P., Eds.; Oxford University Press: Oxford, UK, 2002; pp. 95–281.

5. Barnett, A.J.; Finlay, K.; Beisner, B.E. Functional diversity of crustacean zooplankton communities: Towards a trait-based classification. *Freshw. Biol.* **2007**, *52*, 796–813. [CrossRef]

6. Schleuter, D.; Daufresne, M.; Massol, F.; Argillier, C. A user's guide to functional diversity indices. *Ecol. Monogr.* **2010**, *80*, 469–484. [CrossRef]

7. Cardinale, B.J.; Duffy, J.E.; Gonzalez, A.; Hooper, D.U.; Perrings, C.; Venail, P.; Narwani, A.; Mace, G.M.; Tilman, D.; Wardle, D.A.; et al. Corrigendum: Biodiversity loss and its impact on humanity. *Nature* **2012**, *486*, 59–67.

8. Vogt, R.J.; Beisner, B.E.; Prairie, Y.T. Functional diversity is positively associated with biomass for lake diatoms. *Freshw. Biol.* **2010**, *55*, 1636–1646. [CrossRef]

9. Luoto, T.P.; Nevalainen, L. Climate-forced patterns in midge feeding guilds. *Hydrobiologia* **2014**, *742*, 141–152. [CrossRef]

10. Nevalainen, L.; Luoto, T.P. Relationship between cladoceran (Crustacea) functional diversity and lake trophic gradients. *Funct. Ecol.* **2017**, *31*, 488–498. [CrossRef]

11. Gregory-Eaves, I.; Beisner, B.E. Palaeolimnological insights for biodiversity science: An emerging field. *Freshw. Biol.* **2011**, *56*, 2653–2661. [CrossRef]

12. Nevalainen, L.; Luoto, T.P.; Manca, M.; Weisse, T. A paleolimnological perspective on aquatic biodiversity in Austrian mountain lakes. *Aquat. Sci.* **2014**, *77*, 59–69. [CrossRef]

13. Guilizzoni, P.; Levine, S.N.; Manca, M.; Marchetto, A.; Lami, A.; Ambrosetti, W.; Brauer, A.; Gerli, S.; Carrara, E.A.; Rolla, A.; et al. Ecological effects of multiple stressors on a deep lake (Lago Maggiore, Italy) integrating neo and palaeolimnological approaches. *J. Limnol.* **2012**, *71*, 1–22. [CrossRef]

14. Marchetto, A.; Lami, A.; Musazzi, S.; Massaferro, J.; Langone, L.; Guilizzoni, P. Lake Maggiore (N. Italy) trophic history: Fossil diatom, plant pigments, and chironomids, and comparison with long-term limnological data. *Quat. Int.* **2004**, *113*, 97–110.

15. Morabito, G.; Manca, M. Eutrophication and recovery of the large and deep subalpine Lake Maggiore: Patterns, trends and interactions of planktonic organisms between trophic and climatic forcings. In *Eutrophication: Causes, Economic Implications and Future Challenges*; Environmental Science; Lambert, A., Roux, C., Eds.; NOVA Science Publishers: New York, NY, USA, 2014; pp. 183–214.

16. Manca, M.; Ruggiu, D. Consequences of pelagic food-web changes during a long-term lake oligotrophication process. *Limnol. Oceanogr.* **1998**, *43*, 1368–1373. [CrossRef]

17. Manca, M.; Torretta, B.; Comoli, P.; Amsinck, S.L.; Jeppesen, E. Major changes in trophic dynamics in large, deep sub-alpine lake maggiore from 1940s to 2002: A high resolution comparative palaeo-neolimnological study. *Freshw. Biol.* **2007**, *52*, 2256–2269. [CrossRef]

18. Manca, M.; DeMott, W.R. Response of the invertebrate predator *Bythotrephes* to a climate-linked increase in the duration of a refuge from fish predation. *Limnol. Oceanogr.* **2009**, *54*, 2506–2512. [CrossRef]

19. Szeroczyńska, K.; Sarmaja-Korjonen, K. *Atlas of Subfossil Cladocera from Central and Northern Europe*; Friends of the Lower Vistula Society: Swiecie, Poland, 2007; p. 83.

20. Rao, C.R. Diversity and dissimilarity coefficients: A unified approach. *Theor. Popul. Biol.* **1982**, *21*, 24–43. [CrossRef]

21. Flössner, D.; Krebstiere, C. *Kiemen- und Blattfüßer, Branchiopoda. Fischläuse, Branchiura. Die Tierwelt Deutchlands 60*; Gustav Fischer Verlag: Jena, Germany, 1972; p. 499.

22. Flössner, D. *Die Haplopoda und Cladocera (ohne Bosminidae) Mitteleuropas*; Backhuys Publishers: Leiden, The Netherlands, 2000; p. 428.

23. Šmilauer, P.; Lepš, J. *Multivariate Analysis of Ecological Data Using Canoco 5*; Cambridge University Press: Cambridge, UK, 2014; p. 376.

24. Hammer, Ø.; Harper, D.A.T.; Ryan, P.D. PAST: Paleontological Statistics Software Package for education and data analysis. *Palaeontol. Electron.* **2001**, *4*, 9.

25. Oosterbaan, R.J. SegReg: Segmented Linear Regression with Breakpoint and Confidence Intervals. Available online: https://www.waterlog.info/segreg.htm (accessed on 18 January 2018).

26. Manca, M.M.; Portogallo, M.; Brown, M.E. Shifts in phenology of *Bythotrephes longimanus* and its modern success in Lake Maggiore as a result of changes in climate and trophy. *J. Plankton Res.* **2007**, *29*, 515–525. [CrossRef]

27. Manca, M. Invasions and re-emergences: An analysis of the success of *Bythotrephes* in Lago Maggiore (Italy). *J. Limnol.* **2011**, *70*, 76–82. [CrossRef]

28. Keilty, T.J. A new biological marker layer in the sediments of the great lakes: *Bythotrephes cederstroemi* (schödler) spines. *J. Gt. Lakes Res.* **1988**, *14*, 369–371. [CrossRef]

29. Branstrator, D.K.; Brown, M.E.; Shannon, L.J.; Thabes, M.; Heimgartner, K. Range expansion of *Bythotrephes longimanus* in North America: Evaluating habitat characteristics in the spread of an exotic zooplankter. *Biol. Invasions* **2006**, *8*, 1367–1379. [CrossRef]

30. Hall, R.I.; Yan, N.D. Comparing annual population growth estimates of the exotic invader *Bythotrephes* by using sediment and plankton records. *Limnol. Oceanogr.* **1997**, *42*, 112–120. [CrossRef]

31. Branstrator, D.K.; Beranek, A.E.; Brown, M.E.; Hembre, L.K.; Engstrom, D.R. Colonization dynamics of the invasive predatory cladoceran, *Bythotrephes longimanus*, inferred from sediment records. *Limnol. Oceanogr.* **2017**, *62*, 1096–1110. [CrossRef]

32. Milan, M.; Bigler, C.; Tolotti, M.; Szeroczyńska, K. Effects of long term nutrient and climate variability on subfossil Cladocera in a deep, subalpine lake (Lake Garda, northern Italy). *J. Paleolimnol.* **2017**, *58*, 335–351. [CrossRef]

33. Geller, W.; Müller, H. The Filtration Apparatus of Cladocera: Filter mesh-sizes and their implications on food selectivity. *Oecologia* **1981**, *49*, 316–321. [CrossRef] [PubMed]

34. Vijverberg, J.; Boersma, M. Long-term dynamics of small-bodied and large-bodied cladocerans during the eutrophication of a shallow reservoir, with special attention for *Chydorus sphaericus*. *Hydrobiologia* **1997**, *360*, 233–242. [CrossRef]

35. De Bernardi, R.; Giussani, G.; Manca, M.; Ruggiu, D. Trophic status and the pelagic system in Lago Maggiore. *Hydrobiologia* **1991**, *191*, 1–8. [CrossRef]

36. DeMott, W.R. Feeding selectivities and relative ingestion rates of *Daphnia* and *Bosmina*. *Limnol. Oceanogr.* **1982**, *27*, 518–527. [CrossRef]

37. DeMott, W.R.; Kerfoot, W.C. Competition among cladocerans: Nature of the interaction between *Bosmina* and *Daphnia*. *Ecology* **1982**, *63*, 1949–1966. [CrossRef]

38. Jeppesen, E.; Jensen, J.P.; Amsinck, S.; Landkildehus, F.; Lauridsen, T.; Mitchell, S.F. Reconstructing the historical changes in *Daphnia* mean size and planktivorous fish abundance in lakes from the size of *Daphnia* ephippia in the sediment. *J. Paleolimnol.* **2002**, *27*, 133–143. [CrossRef]

39. Thompson, P.L.; Davies, T.J.; Gonzalez, A. Ecosystem functions across trophic levels are linked to functional and phylogenetic diversity. *PLoS ONE* **2015**, *10*, e0117595. [CrossRef] [PubMed]

40. Amsinck, S.L.; Jeppesen, E.; Landkildehus, F. Inference of past changes in zooplankton community structure and planktivorous fish abundance from sedimentary subfossils—A study of a coastal lake subjected to major fish kill incidents during the past century. *Arch. Hydrobiol.* **2005**, *162*, 363–382. [CrossRef]

Article

Bioassessment of a Drinking Water Reservoir Using Plankton: High Throughput Sequencing vs. Traditional Morphological Method

Wanli Gao [1,2,†], Zhaojin Chen [1,2,†], Yuying Li [1,2,*], Yangdong Pan [2,3], Jingya Zhu [1,2], Shijun Guo [1,2], Lanqun Hu [2,4] and Jin Huang [2,4]

[1] Key Laboratory of Ecological Security for Water Source Region of Middle Route Project of South-North Water Diversion of Henan Province, College of Agricultural Engineering, Nanyang Normal University, Nanyang 473061, Henan, China; highly110@163.com (W.G.); zhaojin_chen@163.com (Z.C.); zhujy15@126.com (J.Z.); 152256558851@163.com (S.G.)

[2] Collaborative Innovation Center of Water Security for Water Source Region of Middle Route Project of South-North Water Diversion of Henan Province, Nanyang Normal University, Nanyang 473061, Henan, China; pany@pdx.edu (Y.P.); hlq65@sina.com (L.H.); nyhbhj@163.com (J.H.)

[3] Department of Environmental Science and Management, Portland State University, Portland, OR 97207, USA

[4] Emergency Centre for Environmental Monitoring of the Canal Head of Middle Route Project of South-North Water Division, Xichuan 474475, Henan, China

* Correspondence: lyying200508@163.com; Tel.: +86-377-6352-5027

† These authors contributed equally to this work.

Received: 25 November 2017; Accepted: 15 January 2018; Published: 18 January 2018

Abstract: Drinking water safety is increasingly perceived as one of the top global environmental issues. Plankton has been commonly used as a bioindicator for water quality in lakes and reservoirs. Recently, DNA sequencing technology has been applied to bioassessment. In this study, we compared the effectiveness of the 16S and 18S rRNA high throughput sequencing method (HTS) and the traditional optical microscopy method (TOM) in the bioassessment of drinking water quality. Five stations reflecting different habitats and hydrological conditions in Danjiangkou Reservoir, one of the largest drinking water reservoirs in Asia, were sampled May 2016. Non-metric multi-dimensional scaling (NMDS) analysis showed that plankton assemblages varied among the stations and the spatial patterns revealed by the two methods were consistent. The correlation between TOM and HTS in a symmetric Procrustes analysis was 0.61, revealing overall good concordance between the two methods. Procrustes analysis also showed that site-specific differences between the two methods varied among the stations. Station Heijizui (H), a site heavily influenced by two tributaries, had the largest difference while station Qushou (Q), a confluence site close to the outlet dam, had the smallest difference between the two methods. Our results show that DNA sequencing has the potential to provide consistent identification of taxa, and reliable bioassessment in a long-term biomonitoring and assessment program for drinking water reservoirs.

Keywords: Danjiangkou Reservoir; plankton; high throughput sequencing; generalized procrustes analysis; bioassessment

1. Introduction

Clean freshwater resources are becoming increasingly scarce globally [1–4]. Due to climate change, economic development and population growth, approximately four billion persons of the world's population are facing severe water scarcity, with nearly half of them living in India and China [1,5]. Drinking water security may substantially hamper the sustainable development of humanity, particularly in developing countries. To better protect and manage diminishing freshwater resources, researchers have developed and tested multiple biological indices for assessing water quality and the ecological integrity of aquatic ecosystems since biological assemblages provide a direct measure of the aquatic ecosystems' conditions [6]. Multimetric indices of fish, macroinvertebrates and periphyton have been effectively used to assess surface water quality in the USA [7]. For example, in an integrated biosurvey as a tool for evaluation of aquatic life use attainment and impairment in Ohio surface waters, Yoder (1991) [8] reported that biological monitoring and assessment indicated that approximately 50% of 645 Ohio stream/river segments were impaired while chemical monitoring and assessment showed no signs of impairment. Biological assessment has become an integral part of water resource protection and management.

One of the key components for a successful bioassessment program is to accurately characterize the composition of biological assemblages. Traditional bioassessment methods rely on specialized analysts to identify taxonomic groups, a time-consuming process. The quality of taxonomic analysis largely depends on analysts training, experience, and interpretation of the taxonomic literature. For instance, several studies on inter-analysts comparison of the diatom identification showed that inconsistency among independent analysts can contribute uncertainty to the bioassessment [9,10]. The inconsistency among the analysts may be more problematic for long-term monitoring programs, particularly with climate change. With the development of DNA sequencing technology [11–14], DNA sequencing has been applied to bioassessment, particularly in freshwater [15–21]. High throughput sequencing (HTS) was developed to characterize biological assemblages in environmental samples. The method is faster in terms of sampling processing, and may become cheaper as the technology improves, and more importantly with the ability to provide more reliable and richer biological information than the traditional morphology-based method [22–24]. Different plankton assemblages as bioindicators were characterized using DNA sequencing including bacterioplankton [18,24–26], and phytoplankton [27–33]. Baird and Hajibabaei (2012) [17] envisioned a new paradigm (i.e., Biomonitoring 2.0) in ecosystem assessment based on a HTS platform, though a complete paradigm shift may require more research.

Several researchers have assessed the effectiveness of water quality assessment using the traditional optic microscopic method and the DNA sequencing method [21,23,34,35]. Due to several limitations, such as incomplete taxonomic coverage in DNA reference libraries and biases related to molecular procedures, the two methods may not generate identical compositional data, particularly at the species level. Despite the difference, biotic indices based on the data generated by the two methods were highly consistent. Multiple researchers suggested that DNA sequencing generates a richer amount of information on biotic diversity with consistent and increased taxonomic resolution and thus has a great potential to improve the effectiveness of current bioassessment. Not surprisingly, most of these studies focused on benthic diatoms in streams and rivers, since diatoms are commonly used as bioindicators in lentic ecosystems. The studies that systematically compare the plankton assemblages characterized by traditional optical microscopy (TOM) and HTS sequencing methods in drinking water reservoirs are still limited.

China is facing challenges in both water quality and quantity. With uneven water distribution, the North China Plain is highly water-stressed, while water resources are relatively more abundant in the south [5,36,37]. Based on the strategic demands for China's regional sustainable development, the Chinese government has launched the South–North Water Diversion (SNWD) Project to transfer water from the Yangtze River and its tributaries to the more arid and industrialized North China Plain [36–38]. As the largest drinking water source in China [38], affecting more than 53 million people in Beijing and other receiving-water regions, the water quality in the Danjiangkou Reservoir, source water for the middle route of the SNWD, is required to be in good quality, be stable long-term, be continuously improved, and be able to adapt to further changes such as climate, and hence the establishment of a continuous ecological monitoring database on the reservoir is particularly important. Such a database can provide scientific data support for best and adaptive management practices.

In this study, we compared the effectiveness of the TOM and HTS methods in assessing water quality in the Danjiangkou Reservoir.

2. Materials and Methods

2.1. Study Area

As one of the largest river impoundments in the Yangtze River basin, the Danjiangkou Reservoir (32°36′~33°48′ N; 110°59′~111°49′ E), with a maximum depth of about 80 m, is located at the juncture of Hubei and Henan provinces of central China. Its drainage area includes the upper Hanjiang and Danjiang rivers, with a total area of 95,000 km^2 (Figure 1). The Danjiangkou Reservoir has a variety of functions, such as flood control, electricity generation, irrigation, shipping and drinking water supply [36–38].

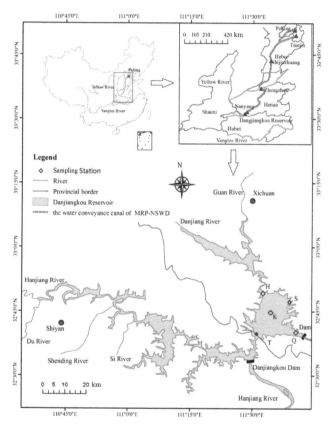

Figure 1. Locations of the five sampling stations in the Danjiangkou Reservoir and the water conveyance canal of the Middle Route Project of South-North Water Division in China. Station codes represent the first letter of their names: Q: Qushou, K: Kuxin, S: Songgan, H: Heijizuo, T: Taizishan.

The reservoir is located in the north sub-tropic monsoon climate region with an annual mean temperature of 15~16 °C and annual precipitation 800~1000 mm, of which 80% is concentrated in the period from May to October. The monthly maximum precipitation is 193.7 mm and the minimum is 31.0 mm. Soil types include alluvial soil, lime concretion black soil, yellow brown loam and purple soil. The watershed around the reservoir is dominated by mountainous areas (85%) and forest coverage (76%) [39].

2.2. Field Sampling

Based on the previous studies [40,41], we selected the five most representative sampling stations in the reservoir. In each station, water samples were collected at the 0~50 cm below the water surface for HTS analysis of planktonic bacterial and eukaryotic assemblages and TOM analysis of the phytoplankton and physico-chemical variables in May 2016 (rainy season). Qushou (Q) station is 100 m upstream of the water outlet dam of the water conveyance canal of the Middle Route Project of SNWD. Heijizuo (H) station is close to the confluence of the Danjiang and Guan rivers, two major tributaries to the reservoir in Henan province. Songgang (S) station is located in a reservoir bay which was influenced by a shipping dock and tourists. Kuxin (K) station is located in the middle of the reservoir while Taizishan (T) station is in the confluence of the two sub-basins of the reservoir (Figure 1). There were three replicates for each sample.

2.3. Morphological Identification of Phytoplankton Assemblages

Phytoplankton samples were preserved immediately with 1% Lugol's solution. The water samples for phytoplankton analysis were stored in a 2 L glass sedimentation utensil. After 48 h sedimentation, each sample was condensed to about 30 mL, and then stored in darkness at 4 °C until analysis. Phytoplankton were identified under a microscope (Nike E200) according to Hu and Wei (2006) [42]. Phytoplankton analysis followed the standard method [43]. Before counting, each sample was gently mixed. Phytoplankton cells were counted in a Fuchs-Rosental counting chamber at $400 \times$ in Z line microscope fields. Algal biovolumes were calculated from measured morphometric characteristics (diameter, length and width). Conversion to biomass was based on 1 mm^3 of algal volume being equivalent to 1 mg of fresh weight biomass. A minimum of 500 individual units were counted with a counting error of less than 10%. Three subsamples were analyzed for each sample.

The Shannon–Wiener index [44] (H') of phytoplankton was used to evaluate water quality. The water quality was classified as severe ($H' = 0{\sim}1$), moderately β ($H' = 1{\sim}2$), moderately α ($H' = 2{\sim}3$) trophic condition and clean condition ($H' > 3$) [45].

2.4. DNA Extraction and Sequencing

Samples were first filtered through a 0.45 μm diameter filter for planktonic eukaryote sequencing and the filtrates were then collected through a 0.22 μm diameter filter for bacterioplankton sequencing. Plankton genomic DNA in the stored filters was extracted using the E.Z.N.A.® Water DNA Kit (OMEGA bio-tek, Norcross, GA, USA), following the manufacturer's instructions. Electrophoresis and Nano Drop ND 2000 (Thermo Scientific, Pittsburgh, PA, USA) were used to estimate the quantity of extracted DNA. The V3-V4 region of bacterial 16S rRNA gene was amplified with 338F and 806R [18,46] and the V5-V7 region of eukaryote 18S rRNA gene was amplified with 0817F and 1196R with sample-identifying barcodes [47]. High throughput sequencing (Illumina MiSeq PE300 platform, Illumina, Inc., CA, USA) was performed by Shanghai Majorbio Bio-Pharm Technology Co., Ltd. (Shanghai, China).

In both cases, the MiSeq sequencing data were processed using the QIIME Pipeline [11]. The operational taxonomic units (OTUs) were determined (at 97% similarity level) using USEARCH. The OTU number of each sample was used to represent taxa richness. Rarefaction curves and a Shannon–Wiener index (H') were generated, and the ACE, Shannon, Simpson, and Chao1 estimators were calculated to compare the richness and diversity of plankton. Taxonomic classification at the phylum and genus levels was performed using the ribosome database project (RDP) algorithm.

2.5. Physico-Chemical Variables

Physico-chemical variables were measured using standard methods [43]. Water temperature (T), pH, electric conductivity (EC), and dissolved oxygen (DO) were measured in situ using a YSI 6920 (YSI Inc., Yellow Springs, OH, USA). Secchi depth (SD) was determined with a 30 cm diameter

Secchi disk. Water samples for chemical analysis were transported to the laboratory within 24 h, stored in a refrigerator at 4 °C, and analyzed within one week after sample collection.

Permanganate index (COD_{Mn}) was calculated using the potassium permanganate index method and chemical oxygen demand (COD) was measured by the potassium dichromate method. Total phosphorus (TP) was determined with acidified molybdate to form reduced phosphor-molybdenum blue and measured spectro-photometrically. Total nitrogen (TN) was assayed with alkaline persulfate digestion and UV spectro-photometry, and ammonia nitrogen (NH_4^+-N) was measured with Nessler's reagent spectro-photometric method. Chlorophyll a (Chla) concentration was estimated spectro-photometrically after extraction in 90% ethanol [43].

The evaluation for the water trophic state was based on the five variables of TN, TP, COD_{Mn}, SD and Chla [48]. Using the linear interpolation method, we converted the value of each environmental variable into the assigned value (E), and then used the arithmetic mean of each assigned value as the trophic state index (EI). EI values between 60~100 indicate hyper-eutrophic conditions, between 50~60 mildly eutrophic conditions and between 20~50 moderately eutrophic conditions [48].

2.6. Data Analysis

Datasets between HTS and TOM were compared at the phylum/genus level. All genera belonging to the concept of phytoplankton, including Cyanobacteria (from the 16S sequencing dataset), Bacillariophyta, Chlorophyta, and other groups of eukaryotic phytoplankton (from the 18S sequencing dataset) were determined from the cleaned HTS datasets. The numbers of OTUs were compared with their corresponding number of phytoplankton species (or genera) found by TOM.

The per-mutational multivariate analysis of variance (PERMANOVA) and analysis of similarity (ANOSIM) were performed using R software with the "vegan" package to assess the significant differences in assemblage structure among the sampling stations. Non-metric multi-dimensional scaling (NMDS) was also conducted with the R function "metaMDS" in the same package, and the first two NMDS axis scores (NMDS I and NMDS II) for each station were used as reduced multi-dimensional data for the plankton assemblages. The relative abundance of biomass data was used for the NMDS analysis. Generalized procrustes analysis (GPA) was used to compare the ordination based on phytoplankton assemblages and DNA barcoding [49]. To assess the differences in the environmental variables among stations, one-way analysis of variance (ANOVA) was applied with statistical significance set prior at $p < 0.05$. All data analyses were performed in R (Ver. 3.4.0, R Development Core Team, 2017).

3. Results

3.1. Evaluation of Water Quality and Trophic Status

The physico-chemical variables indicated that the water quality of the reservoir was overall good as drinking water according to the environmental quality standards of surface water of China (GB38382-2002). However, TN was higher than the Class III surface water standard. TN at H station was 54.08% higher than that of other stations. Both COD_{Mn} and TP met Class II surface water standards while other indicators met Class I surface water standards (Table 1). There were significant differences in the environmental variables among the stations except TP ($p < 0.05$). According to TN, the five stations could be divided into three different groups (i.e., H station, K and T stations, Q and S stations, Table 1).

The mean E value of TN was 60.05. The E values in the H and T stations were higher, over 60.00. The E values of the other four environmental variables were between 20.00~49.05. The mean EI values, ranged between 38.78~41.16, were significantly different among the five stations ($p < 0.05$).

Table 1. Main physico-chemical characteristics of water samples. Station codes represented the first letter of their names: Q: Qushou, K: Kuxin, S: Songgan, H: Heijizuo, T: Taizishang.

Station	pH	SD (m)	DO (mg/L)	TN (mg/L)	NH_4^+-N (mg/L)	TP (mg/L)	COD (mg/L)	COD_{Mn} (mg/L)	Chla (mg/m^3)	EI
Q	8.70 ± 0.009e	3.00 ± 0.000b	7.50 ± 0.035a	0.91 ± 0.003a	0.041 ± 0.005bc	0.02 ± 0a	13.97 ± 0.006d	3.81 ± 0.006e	2.42 ± 0.035bc	41.16 ± 0.003b
K	8.50 ± 0.006d	5.00 ± 0.003e	9.17 ± 0.015d	0.96 ± 0.007b	0.069 ± 0.003d	0.02 ± 0a	11.60 ± 0.145a	3.46 ± 0.006a	2.13 ± 0.015a	38.78 ± 0.002a
S	8.09 ± 0.003a	2.90 ± 0.003a	8.42 ± 0.029c	0.92 ± 0.023ab	0.048 ± 0.001c	0.02 ± 0a	14.37 ± 0.088e	3.74 ± 0.006d	2.40 ± 0.023b	41.13 ± 0.001b
H	8.35 ± 0.003b	4.20 ± 0.003c	8.34 ± 0.020b	1.46 ± 0.020d	0.031 ± 0.001ab	0.02 ± 0a	11.97 ± 0.088b	3.52 ± 0.003b	2.49 ± 0.019cd	41.04 ± 0.003b
T	8.44 ± 0.003c	4.30 ± 0.000d	9.20 ± 0.012d	1.03 ± 0.006c	0.027 ± 0.006a	0.02 ± 0a	12.33 ± 0.033c	3.56 ± 0.015c	2.53 ± 0.003d	40.25 ± 0.000b

Notes: (1) Mean ± standard error; (2) Different lowercase letters in the same column showed that the indicator was significant among stations at $p < 0.05$ level; (3) According to China SL395-2007 evaluation of the water trophic state [48], we converted the value of each environmental variable into the assigned value (E) using the linear interpolation method, and then used the arithmetic mean of each assigned value as the trophic state index (EI).

3.2. Phytoplankton Assemblages Characterized by TOM

A total of 39 taxa were recorded, belonging to five divisions, 17 families and 26 genera. Bacillariophyta, Chlorophyta and Cyanophyta accounted for 51.28%, 23.08% and 15.38% of the total genus/species and 72.53%, 22.15% and 3.78% of the total biomass, respectively (Figure 2). *H'* values varied from the highest of 5.66 at Q station to the lowest of 0.67 at T station (Table A1), suggesting that the reservoir was under a moderate trophic condition.

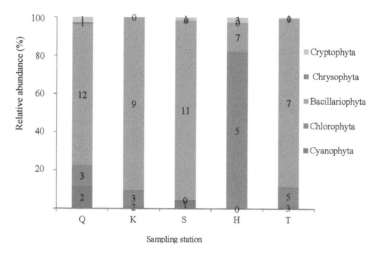

Figure 2. Relative abundance (%) of biomass and taxa number of phytoplankton assemblages by traditional optical microscopy method (TOM). Station codes represented the first letter of their names: Q: Qushou, K: Kuxin, S: Songgan, H: Heijizuo, T: Taizishang.

The NMDS analysis showed that phytoplankton assemblages substantially varied among the five stations (Figure 3a). The NMDS axis I mainly reflected the change of Chlorophyta and Bacillariophyta in the riverine and reservoir areas. The biomass of Chlorophyta, dominated by *Pandorina* sp. (56.34% of the total biomass) at H station, was much higher than that of the other four stations. In contrast, the biomass of Bacillariophyta at H station was the lowest. The NMDS axis II was negatively correlated with the total phytoplankton biomass. *Aulacoseira granulata* (Ehr.) Simonsen accounted for 53.70% and 36.81% of the total biomass in K and T stations, respectively, while *Navicula* sp. accounted for 35.25% and 30.84% in Q and S stations, respectively.

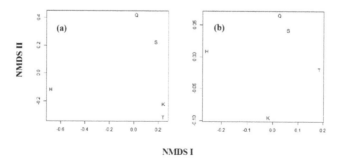

Figure 3. Comparison of the NMDS plots based on phytoplankton (**a**) and DNA barcoding (**b**). Station codes represented the first letter of their names: Q: Qushou, K: Kuxin, S: Songgan, H: Heijizuo, T: Taizishang.

3.3. Plankton Assemblages Characterized by HTS

The HTS results of 16S rRNA and 18S rRNA gene showed that the library coverage and Simpson index of 18S rRNA planktonic eukaryotic assemblages (18S assemblages) were higher than those of 16S rRNA bacterioplankton assemblages (16S assemblages) (Tables 2 and A1, and Figure 4). In contrast, the ACE index, Chao1 index and H' values of the 18S assemblages were lower than those of the 16S assemblages. The OTUs number, ACE index, Chao1 index, H' and Simpson index of the 18S assemblages were significantly different among the stations ($p < 0.05$).

The HTS results showed that the mean bands and OTUs of 16S assemblages of 15 samples were 29,468 and 343 (Table 2 and Figure 4a). Out of 25 phyla of bacteria, Proteobacteria, Bacteroidetes, Actinobacteria, Cyanobacteria and Firmicutes were dominant. The first dominant Proteobacteria was 36.71%, 32.70%, 34.87%, 29.57% and 34.24% of the total OTUs of Q, K, S, H and T stations, and the fourth dominant Cyanobacteria accounted for 7.46%, 10.97%, 11.47%, 9.86% and 8.68%, respectively.

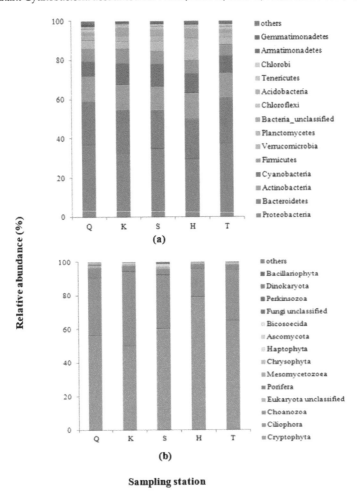

Figure 4. Relative abundance (%) of 16S (**a**) and 18S (**b**) assemblage structures at phylum levels in different stations. Only taxa with 0.05% abundance or more in the total dataset were identified in the legend. Station codes represented the first letter of their names: Q: Qushou, K: Kuxin, S: Songgan, H: Heijizuo, T: Taizishang.

Table 2. Estimation of molecular diversity of 16S and 18S assemblages. Station codes represented the first letter of their names: Q: Qushou, K: Kuxin, S: Songgan, H: Heijizuo, T: Taizishang.

Station	Sequencing Band No.	OTU No.	ACE Index	Chao 1 Index	Library Coverage (%)	Shannon-Wiener Index	Simpson Index
				16S assemblages			
Q	32465 ± 1731a	345 ± 32a	421 ± 29a	421 ± 28a	99.75 ± 0.01a	3.85 ± 0.139a	0.0395 ± 0.007a
K	31068 ± 4516a	334 ± 17a	396 ± 10a	412 ± 10a	99.76 ± 0.05a	4.07 ± 0.070a	0.0293 ± 0.002a
S	27449 ± 3454a	353 ± 37a	420 ± 34a	417 ± 31a	99.70 ± 0.03a	3.97 ± 0.111a	0.0373 ± 0.003a
H	28242 ± 1686a	351 ± 15a	418 ± 12a	414 ± 14a	99.74 ± 0.03a	4.08 ± 0.133a	0.0321 ± 0.005a
T	28116 ± 3473a	332 ± 34a	387 ± 29a	389 ± 24a	99.76 ± 0.03a	4.02 ± 0.135a	0.0332 ± 0.004a
				18S assemblages			
Q	34240 ± 3685a	153 ± 5.49c	168 ± 7.81b	164 ± 5.67b	99.93 ± 0.021a	2.57 ± 0.05c	0.1362 ± 0.005a
K	36042 ± 2763a	104 ± 4.37a	119 ± 5.69a	124 ± 1.15a	99.95 ± 0.004a	2.20 ± 0.02b	0.1785 ± 0.004b
S	36651 ± 1579a	125 ± 4.35b	137 ± 4.70a	137 ± 3.48a	99.95 ± 0.003a	2.47 ± 0.05c	0.1300 ± 0.004a
H	36931 ± 4520a	108 ± 7.31ab	133 ± 13.17a	127 ± 7.21a	99.94 ± 0.014a	2.03 ± 0.01a	0.2355 ± 0.002c
T	33678 ± 3723a	115 ± 3.79ab	128 ± 4.33a	126 ± 2.85a	99.94 ± 0.007a	2.15 ± 0.07ab	0.2041 ± 0.017b

Clustering at 97% Similarity Threshold

Note: (1) Mean ± standard error; (2) Different lowercase letters in the same column indicated that the indicator was significant among stations in $p < 0.05$.

The HTS result of 18S assemblages showed that the mean bands and OTUs were 35,508 and 121 (Table 2). The OTUs of Bikonta, Unikonta and the unclassified accounted for 59.57%, 38.30% and 2.13% of the total OTUs, respectively. There were 20 phyla of planktonic eukaryotes, mainly including Cryptophyta, Ciliophora, Choanomonada, Haptophyta and other groups (Figure 4b). Cryptophyta, Ciliophora and Choanozoa accounted for 97.21% of the total OTUs. Phytoplankton and zooplankton accounted for 63.25% and 35.48% of the total OTUs, respectively. The first dominant Cryptophyta accounted for 56.15%, 49.85%, 60.20%, 79.11% and 64.65% of the total OTUs in Q, K, S, H and T stations, and the second dominant Ciliophora was 34.17%, 44.37%, 32.00%, 16.77% and 30.45%, respectively. Picked phytoplankton included Cryptophyta, Chrysophyta, Haptophyta, Perkinsozoa, Dinokaryota, Bolidomonas, Bacillariophyta, Chlorophyta, Charophyta and Euglenophyta, the first assemblages of which were of 62.00% of the total OTUs.

The NMDS analysis of the DNA sequencing data showed that 16S and 18S assemblages varied among the stations (Figures 3b, 5, 6 and A1). The spatial variation of plankton assemblages among the stations was greater than the measurement error among the replicates of each station (Figures 3b and 5). The difference in bacterioplankton assemblages among the stations was lower than that of eukaryote assemblages (Figures 4 and A1). The bacterioplankton assemblages along the NMDS axis I from left to right may reflect different habitat conditions, and along the NMDS axis II, the differences among bacterioplankton assemblages were less evident (Figure A1a). The OTU percentages of Cyanobacteria_norank and SubsectionI_Family I_norank in H station were higher than those of the other four stations while the OTUs of Synechococcus were much lower. The relative abundance of dominant Cryptomonadales_uncultured along the NMDS axis I from left to right decreased from 43.86% to 14.56%, and along the NMDS axis II from bottom to top, the OTUs of Choanozoa_incertae_sedis gradually dropped from 3.76% to 1.53% (Figure A1b).

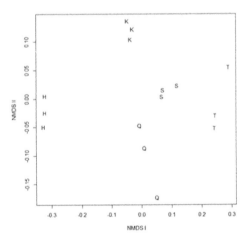

Figure 5. NMDS plot showing both spatial variation among five stations and measurement errors for each station based on the relative abundance of 16S and 18S sequencing barcoding at the genus level. Station codes represented the first letter of their names: Q: Qushou, K: Kuxin, S: Songgan, H: Heijizuo, T: Taizishang. Three identical letters represent three repetitions of each station. There were three repetitions for each sample.

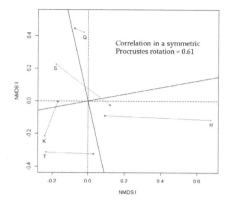

Figure 6. Procrustes plot comparing phytoplankton assemblages and DNA barcoding. Solid lines represented Procrustes residuals from both methods. Station codes represented the first letter of their names: Q: Qushou, K: Kuxin, S: Songgan, H: Heijizuo, T: Taizishang.

3.4. Comparison of Phytoplankton Assemblages by TOM and HTS

The plankton assemblages identified by the two methods were substantially different (Tables 2 and A1; Figures 2–6 and Figure A2). At the phylum level, the number of the phyla identified by HTS was more than that by TOM. However, the genera number with almost the same taxa was similar (Figures 2 and 4). Cyanophyta, Chlorophyta, Bacillariophyta, Chrysophyta and Cryptophyta were detected by both methods with less of the same genera, followed by Microcystis, Anabaena, Navicula and Cryptomonas. Bacillariophyta was dominant with a mean of 51.28% of the total genera/species by TOM while Cryptophyta was dominant with mean 98.27% of the total OTUs by HTS. Dinokaryota, Charophyta, Euglenophyta and the other rare groups were only identified by HTS.

The NMDS analysis showed the substantial spatial difference of the plankton assemblages among the stations using either method (Figures 3, 5, 6 and A1). The plankton assemblages of H station were completely different from those of the other four stations (Figures 3, 5 and 6). The stations could be divided into three group types such as H station, Q and S stations, K and T stations (Figure 3).

Based on the two datasets, GPA was used to compare the difference in assemblages by the two methods. The correlation between TOM and HTS in a symmetric Procrustes rotation was 0.61 (Figure 6), revealing good concordance between the two datasets. The residuals varied among the sampling stations. Q station, located just above the outlet dam, had the smallest residual while H station, located in the confluence of two tributaries with strong river impact, had the largest residual (Figure 6).

4. Discussion

Our study showed that the plankton spatial pattern in the reservoir was consistent using both the TOM and HTS datasets (Figure 3). The reservoir in general has good water quality and thus the spatial differences among the sampled stations may largely reflect the variation of habitat and hydrological conditions. For instance, station H, a strong riverine site located at the confluence of two major tributaries, was consistently different from the rest of the stations in the NMDS plots. Our results are consistent with findings by other researchers [23,50,51]. For instance, both DNA sequencing and TOM analyses generally captured frequency shifts of abundant taxa over the seasonal samples [50]. Vasselon et al. (2017) [23] compared HTS and morphological water quality index values in 33 river sites in Island Mayotte and found that the water quality status was congruent between the two methods. Eiler et al. (2013) [29] reported that DNA-sequencing-derived phytoplankton composition differed significantly among lakes with different trophic status, showing that DNA sequencing could resolve phytoplankton assemblages at a level relevant for ecosystem management.

Procrustes analysis showed that consistency between the two methods varied among the stations in the reservoir (Figure 6). The two methods generated the best agreement at station Q. The station, located just above the outlet dam, may best integrate spatial variability in the reservoir. Both TOM and 18S sRNA results indicated that the sample from this station had the highest taxa richness. In contrast, the least agreement between the two methods was at station H. Located in the downstream of the two tributaries, the station may be heavily influenced by the rivers and consequently its plankton composition with more benthic algae and less true planktonic algae was substantially different from the rest of the stations. The site-specific discrepancy between the two methods may also be due to the incomplete taxonomic coverage in the reference library, particularly for benthic diatoms. For instance, Vasselon et al. (2017) [23] reported that only 13% of the benthic diatom species was shared by the two methods in 33 river sites in Island Mayotte.

It was expected that the plankton assemblages identified by TOM and HTS would be substantially different in terms of their composition. Compared to TOM, HTS can detect a wide variety of plankton, including nanoplankton and rare taxa. Other studies also found similar results in freshwater ecosystems [32,34,50,52]. Xiao et al. (2014) [32] found that the species compositions detected by TOM and 454 HTS did not always match at the taxa level after analyzing 300 weekly water samples over 20 years in Lake Gjersjøen, Norway, a drinking water source. Studies in Lake Tegel, Germany, showed that because the 480 used diatom sequences of the 18S region were generated from world-wide occurrences, only a small number of individuals precisely matched on the species level [52]. A deep-branching taxonomically unclassified cluster was frequently detected by DNA sequencing, but could not be linked to any group identified by microscopy. DNA sequencing allows approximately three orders of magnitude larger SSU rDNA sequencing [16]. Deleting rare species can affect the sensitivity of biotic indices to detect environmental degradation [53]. In the absence of other nuclear markers less susceptible to copy number variation, rDNA-based diversity studies need to be adjusted for confounding effects of copy number variation [50]. Evans et al. (2007) [54] assessed the effectiveness of several genes (cox1, rbcL, 18S and ITS rDNA) to distinguish cryptic species within the model "morphospecies", *Sellaphora pupula agg.*, and found that tree topologies were very similar although support values were generally lower for cox1. To assess the proportional biomass of diatoms and dinoflagellates along the Swedish west marine coast, Godhe et al. (2008) [27] designed two real-time PCR assays with special primers to find that the linear regression of the proportion of SSU rDNA copies of dinoflagellate and diatom origin versus the proportion of dinoflagellate and diatom biovolumes or biomass per liter was significant. Thus, for diatoms, linear regression of the number of small subunit ribosomal DNA (SSU rDNA) copies versus biovolume or biomass per liter was significant, but no such significant correlation was detected in the field samples for dinoflagellates. Exploring the alternative markers (e.g., ~1400 bp of rbcL; 748 bp at the 3_end of rbcL (rbcL-3P); large ribosomal subunit (LSU D2/D3) and UPA) to diatom barcoding (e.g., COI-5P, rbcL-3P) should be used as the primary marker for diatom barcoding, while LSU D2/D3 should be sequenced as a secondary marker to facilitate bioassessment [55].

The high reproducibility and potential for standardization and parallelization makes the high throughput sequencing approach an excellent candidate for the simultaneous monitoring of plankton assemblages in drinking water quality, mainly including both phytoplankton and bacterioplankton. For a long-term biomonitoring and assessment program in the Danjiangkou Reservoir, a critical drinking water source for northern China, our study shows that the DNA barcoding method has great potential. DNA sequencing can provide a rapid and consistent identification of taxa. The next step is to design a specific assay with a specific DNA extraction method and a specific primer for the local flora in the reservoir.

5. Conclusions

In summary, the plankton spatial pattern among the stations in the Danjiangkou Reservoir was consistently detected by both TOM and HTS methods, reflecting the variation of habitat and hydrological conditions in the reservoir. To develop a consistent and accurate long-term ecological monitoring database for drinking water quality, DNA sequencing may serve as a promising alternative.

Acknowledgments: The research was supported by the key Research and Development Program of China (#2016YFC0402204 and 2016YFC0402207), the Major Science and Technology Project for Watershed Pollution Control and Management (#2017ZX07108), the National Natural Science Foundation of China (#3130044, 231400367 and 41601332), the Key Research Project of Colleges and Universities of Henan Province Education Department (#16A210012). We are grateful to the editor, anonymous reviewers and Eugene Foster for their helpful comments and suggestions.

Author Contributions: All the authors contributed equally to this work. Specifically, Wanli Gao and Zhaojin Chen performed the experiments and contributed to writing the paper; Yuying Li and Yangdong Pan conceived and designed the experiments; Jingya Zhu and Shijun Guo did field sampling and analyzed the data. Jin Huang measured the water quality. Lanqun Hu provided the laboratory space for measuring water quality.

Conflicts of Interest: The authors declare no conflict of interest.

Appendix A

Table A1. Shannon-Wiener index (*H'*) of phytoplankton, 16S and 18S assemblages. *H'* values were the mean of the three repetitions of sequencing data. Station codes represented the first letter of their names: H: Heijizuo, K: Kuxin, Q: Qushou, S: Songgan, T: Taizishang.

Station	Phytoplankton Assemblages	16S Assemblages	18S Assemblages
Q	5.66	3.85	2.57
K	1.55	4.07	2.20
S	1.93	3.97	2.47
H	3.85	4.08	2.03
T	0.67	4.02	2.15

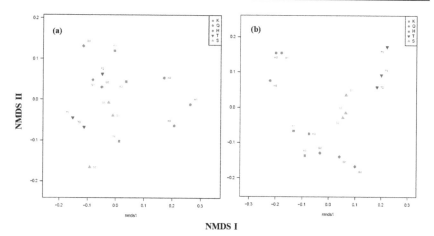

Figure A1. NMDS analysis of 16S (**a**) and 18S (**b**) assemblages structure. Station codes represented the first letter of their names: H: Heijizuo, K: Kuxin, Q: Qushou, S: Songgan, T: Taizishang. The numbers after the station codes indicated the repeating serial numbers.

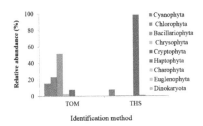

Figure A2. Relative abundance (%) of taxa and OTUs of phytoplankton using traditional optical microscopy method (TOM) and high throughput sequencing method (HTS). OTUs percentage was only that the number of OTUs of one assemblages accounted for the total OTUs number of phytoplankton identified from the 16S and 18S sequencing datasets.

References

1. Vörösmarty, C.J.; Green, P.; Salisbury, J.; Lammers, R.B. Global water resources: Vulnerability from climate change and population growth. *Science* **2000**, *289*, 284–288. [CrossRef] [PubMed]

2. Simonovic, S.P. World water dynamics: Global modeling of water resources. *J. Environ. Manag.* **2002**, *66*, 249–267. [CrossRef]

3. Oki, T.; Kanae, S. Global hydrological cycles and world water resources. *Science* **2006**, *313*, 1068–1072. [CrossRef] [PubMed]

4. Guitton, M.J. The water challenges: Alternative paths to trigger large-scale behavioural shifts. *Lancet Planet. Health* **2017**, *1*, 46–47. [CrossRef]

5. Mekonnen, M.M.; Hoekstra, A.Y. Four billion people facing severe water scarcity. *Sci. Adv.* **2016**, *2*, 1–6. [CrossRef] [PubMed]

6. Karr, J.R. Biological integrity: A long-neglected aspect of water resource management. *Ecol. Appl.* **1991**, *1*, 66–84. [CrossRef] [PubMed]

7. Barbour, M.T.; Gerritsen, J.; Snyder, B.D.; Stribling, J.B. *Rapid Bioassessment Protocols for Use in Streams and Wadeable Rivers: Periphyton, Benthic Macroinvertebrates and Fish*, 2nd ed.; Environmental Protection Agency, Office of Water: Washington, DC, USA, 1999; pp. 23–178.

8. Yoder, C. The integrated biosurvey as a tool for evaluation of aquatic life use attainment in Ohio surface waters. In *Biological Criteria: Research and Regulation*; Environmental Protection Agency: Washington, DC, USA, 1991; pp. 110–122.

9. Prygiel, J.; Carpentier, P.; Almeida, S.; Coste, M.; Druart, J.C.; Ector, L.; Guillard, D.; Honoré, M.A.; Iserentant, R.; Ledeganck, P.; et al. Determination of the biological diatom index (IBD NF T 90–354): Results of an intercomparison exercise. *J. Appl. Phycol.* **2002**, *14*, 27–39. [CrossRef]

10. Kelly, M.; Bennett, C.; Coste, M.; Delgado, C.; Delmas, F.; Denys, L.; Ector, L.; Fauville, C.; Ferréol, M.; Golub, M.; et al. A comparison of national approaches to setting ecological status boundaries in phytobenthos assessment for the European Water Framework Directive: Results of an intercalibration exercise. *Hydrobiologia* **2009**, *621*, 169–182. [CrossRef]

11. Caporaso, J.G.; Kuczynski, J.; Stombaugh, J.; Bittinger, K.; Bushman, F.D.; Costello, E.K.; Fierer, N.; Peña, A.G.; Goodrich, J.K.; Gordon, J.I.; et al. QIIME allows analysis of high-throughput community sequencing data. *Nat. Methods* **2010**, *7*, 335. [CrossRef] [PubMed]

12. van Dijk, E.L.; Auger, H.; Jaszczyszyn, Y.; Thermes, C. Ten years of next-generation sequencing technology. *Trends Genet.* **2014**, *30*, 418–426. [CrossRef] [PubMed]

13. Esling, P.; Lejzerowicz, F.; Pawlowski, J. Accurate multiplexing and filtering for high-throughput amplicon-sequencing. *Nucleic Acids Res.* **2015**, *43*, 2513–2524. [CrossRef] [PubMed]

14. Evans, K.M.; Mann, D.G. A Proposed Protocol for Nomenclaturally Effective DNA Barcoding of Microalgae. *Phycologia* **2009**, *48*, 70–74. [CrossRef]

15. Dorigo, U.; Volatier, L.; Humbert, J.F. Molecular approaches to the assessment of biodiversity in aquatic microbial communities. *Water Res.* **2005**, *39*, 2207–2218. [CrossRef] [PubMed]

16. Creer, S. Second-generation sequencing derived insights into the temporal biodiversity dynamics of freshwater protists. *Mol. Ecol.* **2010**, *19*, 2829–2831. [CrossRef] [PubMed]

17. Baird, D.J.; Hajibabaei, M. Biomonitoring 2.0: A new paradigm in ecosystem assessment made possible by next-generation DNA sequencing. *Mol. Ecol.* **2012**, *21*, 2039–2044. [CrossRef] [PubMed]

18. Zhang, Y. Microbial Community Dynamics and Assembly: Drinking Water Treatment and Distribution. Ph.D. Thesis, University of Tennessee, Knoxville, TN, USA, 2012.

19. Stein, E.D.; Martinez, M.C.; Stiles, S.; Miller, P.E.; Zakharov, E.V. Is DNA Barcoding Actually Cheaper and Faster than Traditional Morphological Methods: Results from a Survey of Freshwater Bioassessment Efforts in the United States? *PLoS ONE* **2014**, *9*, e95525. [CrossRef] [PubMed]

20. Stein, E.D.; White, B.P.; Mazor, R.D.; Jackson, J.K.; Battle, J.M.; Miller, P.E.; Pilgrim, E.M.; Sweeney, B.W. Does DNA Barcoding Improve Performance of Traditional Stream Bioassessment Metrics? *Freshw. Sci.* **2013**, *33*, 302–311. [CrossRef]

21. Visco, J.A.; Apothéloz-Perret-Gentil, L.; Cordonier, A.; Esling, P.; Pillet, L.; Pawlowski, J. Environmental Monitoring: Inferring the Diatom Index from Next-Generation Sequencing Data. *Environ. Sci. Technol.* **2015**, *49*, 7597–7605. [CrossRef] [PubMed]

22. Keck, F.; Vasselon, V.; Tapolczai, K.; Rimet, F.; Bouchez, A. Freshwater biomonitoring in the Information Age. *Front. Ecol. Environ.* **2017**, *15*, 266–274. [CrossRef]

23. Vasselon, V.; Rimet, F.; Tapolczai, K.; Bouchez, A. Assessing ecological status with diatoms DNA metabarcoding: Scaling-up on a WFD monitoring network (Mayotte island, France). *Ecol. Ind.* **2017**, *82*, 1–12. [CrossRef]

24. Zhang, J.; Zhu, C.; Guan, R.; Xiong, Z.; Zhang, W.; Shi, J.; Sheng, Y.; Zhu, B.; Tu, J.; Ge, Q.; et al. Microbial profiles of a drinking water resource based on different 16S rRNA V regions during a heavy cyanobacterial bloom in Lake Taihu, China. *Environ. Sci. Pollut. Res.* **2017**, *24*, 12796–12808. [CrossRef] [PubMed]

25. Glöckner, F.O.; Zaichikov, E.; Belkova, N.; Denissova, L.; Pernthaler, J.; Pernthaler, A.; Amann, R. Comparative 16S rRNA analysis of lake bacterioplankton reveals globally distributed phylogenetic clusters including an abundant group of Actinobacteria. *Appl. Environ. Microbiol.* **2000**, *66*, 5053–5065. [CrossRef] [PubMed]

26. Zhao, D.; Shen, F.; Zeng, J.; Huang, R.; Yu, Z.; Wu, Q.L. Network analysis reveals seasonal variation of co-occurrence correlations between Cyanobacteria and other bacterioplankton. *Sci. Total Environ.* **2016**, *573*, 817–825. [CrossRef] [PubMed]

27. Godhe, A.; Asplund, M.E.; Härnström, K.; Saravanan, V.; Tyagi, A.; Karunasagar, I. Quantification of diatom and dinoflagellate biomasses in coastal marine seawater samples by real-time PCR. *Appl. Environ. Microbiol.* **2008**, *74*, 7174–7182. [CrossRef] [PubMed]

28. Monchy, S.; Grattepanche, J.D.; Breton, E.; Meloni, D.; Sanciu, G.; Chabé, M.; Delhaes, L.; Viscogliosi, E.; Sime-Ngando, T.; Christaki, U. Microplanktonic community structure in a coastal system relative to a phaeocystis bloom inferred from morphological and tag pyrosequencing methods. *PLoS ONE* **2012**, *7*, e39924. [CrossRef] [PubMed]

29. Eiler, A.; Drakare, S.; Bertilsson, S.; Pernthaler, J.; Peura, S.; Rofner, C.; Simek, K.; Yang, Y.; Znachor, P.; Lindström, E.S. Unveiling Distribution Patterns of Freshwater Phytoplankton by a Next Generation Sequencing Based Approach. *PLoS ONE* **2013**, *8*, e53516. [CrossRef] [PubMed]

30. Alemzadeh, E.; Haddad, R.; Ahmadi, A.R. Phytoplanktons and DNA barcoding: Characterization and molecular analysis of phytoplanktons on the Persian Gulf. *Iran. J. Microbiol.* **2014**, *6*, 296–302. [PubMed]

31. Bazin, P.; Jouenne, F.; Friedl, T.; Deton-Cabanillas, A.F.; Le Roy, B.; Véron, B. Phytoplankton diversity and community composition along the estuarine gradient of a temperate macrotidal ecosystem: Combined morphological and molecular approaches. *PLoS ONE* **2014**, *9*, e94110.

32. Xiao, X.; Sogge, H.; Lagesen, K.; Tooming-Klunderud, A.; Jakobsen, K.S.; Rohrlack, T. Use of high throughput sequencing and light microscopy show contrasting results in a study of phytoplankton occurrence in a freshwater environment. *PLoS ONE* **2014**, *9*, e101560. [CrossRef] [PubMed]

33. Taylor, J.D.; Cottingham, S.D.; Billinge, J.; Cunliffe, M. Seasonal microbial community dynamics correlate with phytoplankton-derived polysaccharides in surface coastal waters. *ISME J.* **2013**, *8*, 245. [CrossRef] [PubMed]

34. Zimmermann, J.; Abarca, N.; Enk, N.; Skibbe, O.; Kusber, W.H.; Jahn, R. Taxonomic reference libraries for environmental barcoding: A best practice example from diatom research. *PLoS ONE* **2014**, *9*, e108793. [CrossRef] [PubMed]

35. Kermarrec, L.; Franc, A.; Rimet, F.; Chaumeil, P.; Frigerio, J.M.; Humbert, J.F.; Bouchez, A. A next-generation sequencing approach to river biomonitoring using benthic diatoms. *Freshw. Sci.* **2014**, *33*, 349–363. [CrossRef]

36. Berkoff, J. China: The south-north water transfer project-is it justified? *Water Policy* **2003**, *5*, 1–28.

37. Shao, X.; Wang, H.; Wang, Z. Interbasin transfer projects and their implications: A China case study. *Int. J. River Basin Manag.* **2003**, *1*, 5–14. [CrossRef]

38. Zhang, Q. The south-to-north water transfer project of China: Environmental implications and monitoring strategy. *J. Am. Water Resour. Assoc.* **2009**, *45*, 1238–1247. [CrossRef]

39. Shen, H.; Cai, Q.; Zhang, M. Spatial gradient and seasonal variation of trophic status in a large water supply reservoir for the south-to-north water diversion project, China. *J. Freshw. Ecol.* **2015**, *30*, 249–261. [CrossRef]

40. Li, Y.Y.; Gao, W.L.; Li, J.F.; Wen, Z.Z.; Liu, H.; Hu, L.Q.; Zhang, N.Q.; Cheng, X. Spatio-temporal distribution of phytoplankton and trophic status in the water resource area of the middle route of China's south to north water transfer project. *Chin. J. Ecol.* **2008**, *27*, 14–22. (In Chinese with English Abstract)

41. Chen, Z.; Ding, C.; Zhu, J.; Li, B.; Huang, J.; Du, Z.; Wang, Y.; Li, Y. Community structure and influencing factors of bacterioplankton during low water periods in Danjiangkou Reservoir. *China Environ. Sci.* **2017**, *37*, 336–344, (In Chinese with English Abstract). [CrossRef] [PubMed]

42. Hu, H.J.; Wei, Y.X. *The Freshwater Algae of China: Systematics, Taxonomy and Ecology*; Science and Technology Press: Beijing, China, 2006. (In Chinese)

43. Chinese State Environment Protection Bureau (CSEPB). *Water and Wastewater Monitoring Analysis Methods*, 4th ed.; Chinese Environment Science Press: Beijing, China, 2006; pp. 20–345. (In Chinese)

44. Shannon, C.E.; Wiener, W. *The Mathematical Theory of Communication*; University of Illinois Press: Urbana, IL, USA, 1949; p. 117.

45. Jin, X.C.; Tu, Q.Y. *The Standard Methods for Observation and Analysis of Lake Eutrophication*, 2rd ed.; China Environmental Science Press: Beijing, China, 1990; pp. 10–167. (In Chinese)

46. Lee, C.K.; Barbier, B.A.; Bottos, E.M.; McDonald, I.R.; Cary, S.C. The Inter-Valley Soil Comparative Survey: The ecology of Dry Valley edaphic microbial communities. *ISME J.* **2011**, *6*, 1046. [CrossRef] [PubMed]

47. Wang, Y.; Zheng, Y.; Wang, X.; Wei, X.; Wei, J. Lichen-associated fungal community in Hypogymnia hypotrypa (Parmeliaceae, Ascomycota) affected by geographic distribution and altitude. *Front. Microbiol.* **2016**, *7*, 1231. [CrossRef] [PubMed]

48. Water Resources Department of the People's Republic of China. *SL395-2007 Technological Regulations for Quality Assessment of Surface Water Resources in China*; China Water Conservancy and Hydropower Press: Beijing, China, 2007; pp. 1–5. (In Chinese)

49. Peres-Neto, P.R.; Jackson, D.A. How well do multivariate data sets match? The advantages of a Procrustean superimposition approach over the Mantel test. *Oecologia* **2001**, *129*, 169–178. [CrossRef] [PubMed]

50. Medinger, R.; Nolte, V.; Pandey, R.V.; Jost, S.; OttenwÄLder, B.; SchlÖTterer, C.; Boenigk, J. Diversity in a hidden world: Potential and limitation of next-generation sequencing for surveys of molecular diversity of eukaryotic microorganisms. *Mol. Ecol.* **2010**, *19*, 32–40. [CrossRef] [PubMed]

51. Nolte, V.; Pandey, R.V.; Jost, S.; Medinger, R.; OttenwÄLder, B.; Boenigk, J.; SchlÖTterer, C. Contrasting seasonal niche separation between rare and abundant taxa conceals the extent of protist diversity. *Mol. Ecol.* **2010**, *19*, 2908–2915. [CrossRef] [PubMed]

52. Jahn, R.; Zetzsche, H.; Reinhardt, R.; Gemeinholzer, B. *Diatoms and DNA Barcoding: A Pilot Study on an Environmental Sample*; Kusber, W.H., Jahn, R., Eds.; Botanic Garden and Botanical Museum Berlin: Dahlem, Germany; Freie Universität Berlin: Berlin, Germany, 2007; pp. 63–68.

53. Cao, Y.; Williams, D.D.; Williams, N.E. How important are rare species in aquatic community ecology and bioassessment? *Limnol. Oceanogr.* **1998**, *43*, 1403–1409. [CrossRef]

54. Evans, K.M.; Wortley, A.H.; Mann, D.G. An assessment of potential diatom "barcode" genes (cox1, rbcL, 18S and ITS rDNA) and their effectiveness in determining relationships in Sellaphora (Bacillariophyta). *Protist* **2007**, *158*, 349–364. [CrossRef] [PubMed]

55. Hamsher, S.E.; Evans, K.M.; Mann, D.G.; Poulíčková, A.; Saunders, G.W. Barcoding diatoms: Exploring alternatives to COI-5P. *Protist* **2011**, *162*, 405–422. [CrossRef] [PubMed]

 *water*

Article

Interspecific Relationship and Ecological Requirements of Two Potentially Harmful Cyanobacteria in a Deep South-Alpine Lake (L. Iseo, I)

Veronica Nava, Martina Patelli, Valentina Soler and Barbara Leoni *

Department of Earth and Environmental Sciences, University of Milano-Bicocca, Piazza Della Scienza 1, 20126 Milano, Italy; v.nava15@campus.unimib.it (V.N.); m.patelli3@campus.unimib.it (M.P.); valentina.soler@unimib.it (V.S.)
* Correspondence: barbara.leoni@unimib.it; Tel.: +39-026-448-2712

Received: 17 November 2017; Accepted: 14 December 2017; Published: 19 December 2017

Abstract: In Lake Iseo (Lombardia, Italy), the predominant species in the cyanobacterial taxa was *Planktothrix rubescens*. However, since 2014, the presence of an allochthonous Cyanobacteria, *Tychonema bourrellyi*, able to produce consistent biomasses and harmful toxins, was detected. The causes of this expansion are poorly understood. Many studies have linked the development of allochthonous Cyanobacteria populations with climate change. This study shows the spatio-temporal dynamics, the ecological requirements, and the interspecific relationship of *P. rubescens* and *T. bourrellyi*. Samples were collected monthly in 2016 at six different depths in the water column; 20 chemico-physical characteristics were measured; and Cyanobacteria density, morphology, and biovolume were evaluated. The results allowed a comparison of the spatial pattern of the two species, which showed a greater distribution at a depth of 10–20 m, and their seasonal dynamics. Both Cyanobacteria were present throughout the year, with the greatest abundance during the period from March to May. A temporal shift was observed in their development, linked to different capacities for overcoming winter and mixing periods. Principal Component Analysis, performed on 20 observations (4 months × 5 depths), highlighted the important role of the stability of the water column in determining *T. bourrellyi* settlement in Lake Iseo and the role of solar radiation in spring population development.

Keywords: Cyanobacteria; colonization; *Tychonema bourrellyi*; *Planktothrix rubescens*; deep lake; stability; PCA

1. Introduction

Cyanobacteria are the most ancient phototrophs on the earth. As a result of their long evolutionary history (~2 by) [1,2], and due to their high ecological plasticity, Cyanobacteria are able to adapt to geochemical and climatic changes [3,4]. Consequently, these organisms are also capable of adapting to anthropogenic modifications of aquatic environments [5], such as enhanced nutrient loading or increasing temperatures. The literature has reported that climate change is likely to stimulate the development of harmful Cyanobacteria (e.g., [6–14]). Regional and global warming, with an associated increase in temperature and variability in rainfall patterns, causes changes in nutrients, sediment delivery, sediment-water exchange and metabolism, water residence time, and vertical stratification [6,12]. Therefore, ecosystems are subjected to alterations that affect the biotic community, modifying the species composition with the invasion and the establishment of allochthonous populations of Cyanobacteria in freshwater ecosystems [4,15,16].

Increasing focus is being directed towards harmful Cyanobacteria in aquatic ecosystems, as these organisms are capable of forming consistent blooms (cyanoHABs) affecting use, safety, and sustainability of water resources, with considerable ecological and socioeconomic costs [1,17,18]. Cyanobacteria blooms represent an appreciable threat not only to human health through exposure to toxic compounds, such as via drinking water, aquaculture, and recreation [18,19], but also for the biological community, because the proliferation of these organisms can affect the functioning of the ecosystem [11,17]. Nevertheless, the impact of cyanotoxins on aquatic ecosystems remains poorly understood [20], so acquiring more information about the toxicological features and the autoecology of the species able to produce cyanotoxins is important. These information would help increase understanding of the environmental drivers that promote the proliferation of harmful Cyanobacteria, and consequently provide further information for water managers and policy makers.

Studies on the phytoplanktonic community and Cyanobacteria species have been performed for over 30 years in the great south-Alpine lakes in Italy (i.e., L. Maggiore, L. Como, L. Iseo, L. Garda) ([21–28]). Starting in 2014, the colonization and the development of *Tychonema bourrellyi* (J. W. G. Lund) [29], a Cyanobacteria able to produce anatoxin-A (ATX), were documented [3,4]. The extent and the causes of the expansion of this species in deep holo-oligomictic lakes (i.e., large lakes south of the Alps) remain to be explored [4]; information about its ecology and physiology are limited [30].

The aims of this paper were to understand how this allochthonous species affects the current cyanobacterial community, to determine which environmental factors favored the invasion, and to investigate the direct or indirect relationships between environmental changes (i.e., climate) and the development of potentially harmful Cyanobacteria. These goals were achieved by analyzing phytoplanktonic populations in Lake Iseo, a south-Alpine deep lake, in which one of the predominant Oscillatoriales over the past 15 years was *Planktothrix rubescens* (De Candolle ex Gomont) [3,22,23,29,31]. The main objectives were: (1) to highlight and compare spatio-temporal population dynamics of *P. rubescens* and *T. bourrellyi*, analyzing their maximum and minimum seasonal abundance, along with their distribution along the water column, to determine if the presence of an allochthonous species could modify the dynamics of the preexisting cyanobacterial community; (2) to evaluate lacustrine chemico-physical parameters that could affect their development. We supposed that the environmental changes that occurred in recent decades in Lake Iseo, driven by a climatic force acting at a regional scale [32–34], affected the phytoplanktonic community. In particular, the documented increases in water temperature and water column stability [35] could promote the colonization and development of *T. bourrellyi* in Lake Iseo.

2. Materials and Methods

2.1. Study Area

Lake Iseo is a deep perialpine lake, the fifth-largest Italian lake in terms of volume (7.6 km^3). The surface area is 61.8 km^2, the maximum depth is 258 m, and the mean depth is 124 m [22,36,37]. The lake, which is included in the LTER network (Site LTER_EU_IT_008—"Southern Alpine Lakes"; http://www.lter-europe.net), is situated in Northern Italy at the end of a prealpine valley, Val Camonica (Figure 1).

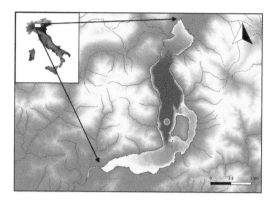

Figure 1. Location and bathymetry of Lake Iseo. The red dot indicates the position of the sampling point (modified [38]).

Lake Iseo should be classified as "warm monomictic", since winter water temperatures never drop below 4 °C. However, due to the great depth of the lake, winter mixing occurs only during cold and windy winters. Over the last twenty years, complete winter mixing occurred only in 2005 and 2006, so the lake can now be regarded as holo-oligomictic [21,22,36,39,40]. Except for these two years, the spring mixing depth from 1998 to 2016 in Lake Iseo ranged between 30 and 150 m, and in 2016 it reached about 50 m [35,36]. The extent of the mixing depth was determined by analyzing the homogeneity of the depth profiles of physical and chemical data recorded along the water column [35].

Nowadays, the increase in nutrient loading has brought Lake Iseo to a meso-eutrophic condition, with an average concentration of total phosphorus of around 80–90 μg/L [21,22,36].

2.2. Sampling, Field Measurements, and Laboratory Analyses

Samplings for the analysis of the two Cyanobacteria were carried out in Lake Iseo from January to December 2016. Samples were collected at the deepest point of the lake (45°43′11″ N; 10°03′46″ E) [22], at an almost monthly frequency, using a Niskin bottle, along the water column at six different depths (0, 10, 20, 30, 40, 60 m), collecting over all 72 samples. These depths were chosen in consideration of the extent of the euphotic zone in Lake Iseo.

An algae count was performed on fresh material, due to the difficulty of discriminating the filaments of *T. bourrellyi* and *P. rubescens* in samples fixed with Lugol solution [4]. From each sample, 10 mL was vacuum filtered onto 5 μm pore-sized Whatman Cyclopore transparent polycarbonate membrane filters (25 mm filter diameter) [4,41]. Filtration was made at low pressure to avoid damaging the filaments [4]. The filters were transferred on a microscopic slide and, employing a microscope (100× magnification), discrimination and quantification of the two species were carried out, analyzing the filaments in the whole area of the filters. The average length and average width of the cells were obtained by measuring at least 20 randomly selected sample filaments. These parameters allowed the estimation of their density (cell/L) and biovolume (mm^3/m^3), assuming cylindrical cell shape for both the taxa [42,43].

In the laboratory, determinations of total phosphorus, soluble reactive phosphorus, total nitrogen, nitrate nitrogen, ammonium nitrogen, conductivity, and alkalinity were carried out using standard methods, in accordance with APHA-AWWA-WEF [44]. Dissolved oxygen concentration, pH, and temperature were, however, detected in situ, instead, with a portable underwater multiparameter probe (WTW multi3432). Cations and anions were measured using the Ion Chromatography (Thermo Scientific™ Dionex™, Waltham, MA, USA).

The water transparency (z_s) was detected monthly using the Secchi disk, and the depth of the euphotic zone (z_{eu}) was estimated as $z_{eu} = 4.8 \cdot z_s^{0.68}$ [45]. The thermocline depth was calculated monthly using the package 'rLakeAnalyzer' in R 3.4.1. [46], where the minimum density gradient was set by default to 0.1 kg/m^3/m. The penetration of solar radiation into the water column was obtained from the values of superficial global radiation (W/m^2) provided by ARPA Lombardia, with measurements made at the meteorological station of Costa Volpino (192 m a.s.l.), located ca. 1 km away from the northern border of the lake. Stability was, moreover, calculated monthly as the Brunt-Väisälä frequency (s^{-2}), using the package 'rLakeAnalyzer' [46,47].

2.3. Statistical Analyses

In order to highlight the ecological requirements of *P. rubescens* and *T. bourrellyi*, the relationship among selected chemico-physical parameters and biovolume of the two Cyanobacteria was analyzed using Principal Component Analysis (PCA) based on the correlation matrix [48,49].

Twenty chemico-physical parameters were selected for the PCA: water temperature "WT" (°C), water density "WD" (kg/m^3), pH, alkalinity "ALK" (meq/L), electrical conductivity "EC" (µS/cm), total phosphorus "TP" (µg/L), total nitrogen "TN" (µg/L), TP:TN ratio "TP.TN", soluble reactive phosphorus "SRP" (µg/L), ammonium nitrogen "NH4" (µg/L), nitrate nitrogen "NO3" (µg/L), silicates "Si" (µg/L), chloride "Cl" (mg/L), sulphates "SO4" (mg/L), calcium "Ca" (mg/L), magnesium "Mg" (mg/L), sodium "Na" (mg/L), potassium "K" (mg/L), light radiation "RAD" (W/m), and Brunt-Väisälä frequency "STABILITY" (s^{-2}).

All data were centered (mean value = 0) and scaled (variance = 1) to allow comparison among parameters. The normality distribution was checked using the Shapiro-Wilk test [50], and variables without a normal distribution were log$_{10}$(x + 1) transformed. The species biovolumes were Hellinger-transformed, in order to reduce the importance of large abundances [49,51,52].

A preliminary principal component analysis was performed, including all 72 observations, in order to highlight the most relevant parameters to focus on. A final PCA was performed considering only the months and the depths possessing the strongest relationships between the Cyanobacteria and the environmental features. The more relevant relationships between *T. bourrellyi* and chemico-physical parameters, showed by PCA, were tested with a linear regression analysis. All statistical analyses were performed by different packages in R 3.4.1. [53].

3. Results

3.1. Thermal Regime and Water Stability

From January to December 2016, temperature values in the first 60 m of the water column of Lake Iseo were between 6.5 and 23.8 °C (Figure 2).

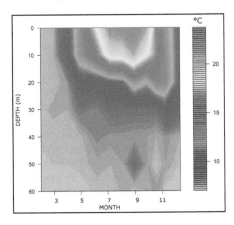

Figure 2. Heatmap of water temperature in Lake Iseo from January to December 2016.

The warmest temperature values were detected in the first 10–20 m during summer, especially in July and August. Below a depth of 60 m, the water temperature was quite constant (6.8 ± 0.7 °C (±SEM)). The lowest water temperature difference in the water column was detected in January and February, with a delta from the top (0 m) to the bottom (245 m) of 0.80 °C. In this period, we assumed water mixing only occurred up to a depth of 50 m by analyzing the chemical parameters along the water column. Starting mid-March, we observed a progressively more pronounced difference in the water column temperature among the superficial and the deeper layers, with an enhanced stratification from May to September. The depth of the thermocline (z_{therm}) ranged from 13 to 15 m.

During 2016, the value of water transparency in Lake Iseo ranged from 2.4 (June) to 10 m (January). In general terms, a greater transparency was observed during the winter months due to the lower primary production compared to summer months. The euphotic depth, ranging between 9 (June) and 23 m (January), was attained using the Secchi disk (z_s). The surface radiation had values of between 40 and 293 W/m^2, with an extinction coefficient in the range of 0.2 to 0.5.

The Brunt-Väisälä frequency (STABILITY) ranged between -6.7×10^{-5} and 1.43×10^{-3} s^{-2}. The values detected at the thermocline ranged between 1.5×10^{-5} and 1.4×10^{-3} s^{-2} and reached maximum values during the summer months, whereas the minimum values were observed during winter [20,54].

3.2. Chemical Characteristics

Dissolved oxygen (DO) was higher in shallower layers. In Lake Iseo, below a depth of 50 m, the oxygen concentration was always critical, due to the high lake productivity, as well as the diminished frequency of deep circulation recorded in recent years. During February and March 2016, a dissolved oxygen recharge in the 40–60 m layers was observed, with a rapid subsequent decrease. During August and September, the metalimnetic oxygen concentration was higher than the levels in the hypolimnion and epilimnion, displaying a positive heterograde curve (Figure 3a).

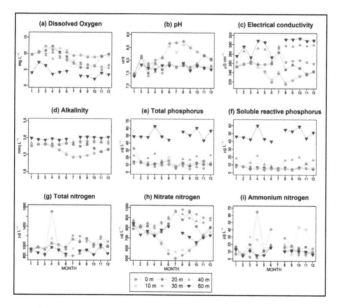

Figure 3. Temporal development of the principal chemico-physical features analyzed at six depths: (**a**) Dissolved oxygen, DO (mg/L); (**b**) pH; (**c**) Electrical conductivity, EC (µS/cm); (**d**) Alkalinity, ALK (meq/L); (**e**) Total phosphorus, TP (µg/L); (**f**) Soluble reactive phosphorus, SRP (µg/L); (**g**) Total nitrogen, TN (µg/L); (**h**) Nitrate nitrogen, NO$_3^-$ (µg/L); (**i**) Ammonium nitrogen, NH$_4^+$ (µg/L).

pH, measured at 20 °C, showed a similar pattern as oxygen, with a higher value in the superficial layers, where the photosynthetic activity of autotrophic organisms is concentrated, and lower values in deeper layers, where heterotrophs produce carbon dioxide (CO_2), moving the pH toward neutral values. pH values of between 7.1 and 8.7 were recorded, with higher superficial values in June, July, and August, and with the lowest value being recorded in January (Figure 3b).

During the whole period analyzed, alkalinity (ALK) and electric conductivity (EC) were in the range of 1.4 to 2.3 meq/L and 216 to 320 µS/L, respectively (Figure 3c,d).

Total phosphorus (TP) and soluble reactive phosphorus (SRP) showed a similar trend. In epilimnetic layers, the concentration was 12.9 ± 0.7 µg/L and 6.33 ± 0.28 µg/L, respectively for TP and SRP. At a depth of 60 m, where the primary production is low and the microbial loop is well developed, the concentration was 51.9 ± 1.9 µg/L for TP and 49.0 ± 2.1 µg/L for SRP (Figure 3e,f).

In the first 60 m, total nitrogen (TN) was between 800 and 1713 µg/L, with the peak value topside in April and the lowest value in June at 0 m (Figure 3g). The maximum mean value along the water column, from the top to a depth of 60 m, was recorded in April (1153 µg/L), whereas the minimum value was recorded in June (887 µg/L). Nitrogen nitrate ($N\text{-}NO_3^-$) showed lower values at 0 and 10 m, whereas the greatest values were detected at 30 and 40 m, within the range of 505 to 934 µg/L (Figure 3h). Nitrogen ammonium ($N\text{-}NH_4^+$) ranged between 2 and 65 µg/L, with the maximum value recorded in April at 0 m and the minimum value also in April at 30 m (Figure 3i).

Silicates had values between 56 and 1524 µg/L, with the greatest value recorded at 60 m. The silicate concentrations did not differ between the various sampling months, remaining quite constant. Chloride, sodium, and potassium ions displayed similar values throughout the year and along the water column, with values of 3.07 ± 0.02, 2.81 ± 0.01, and 1.29 ± 0.01 mg/L, respectively. For magnesium and sulphates, with respective mean values of 7.45 ± 0.35 and 18.5 ± 0.2 mg/L, a variation in seasonal and spatial values was not observed. Calcium ions showed similar values in the different months at various depths, with a mean value of 41.8 ± 0.39 mg/L, although a slight decrease in the superficial layers (i.e., 0–10 m) was seen from May to October (35.3 ± 0.53 mg/L).

3.3. Seasonal and Spatial Dynamics of P. rubescens and T. bourrellyi

During the observation period of January to December 2016, both Cyanobacteria were always present in the lake. The mean species biovolume (from top to 60 m) reached its maximum in April, with a value of 585 ± 21 mm³/m³ for *P. rubescens* and 162 ± 72 mm³/m³ for *T. bourrellyi*. The minimum mean biovolume was recorded in September, with a value of 1.8 ± 0.7 mm³/m³ for *P. rubescens* and 4.9 ± 1.5 mm³/m³ for *T. bourrellyi* (Figure 4).

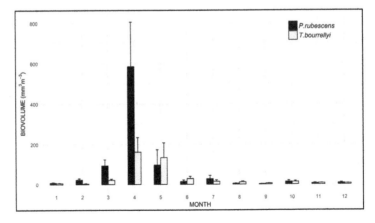

Figure 4. Monthly mean biovolume ± SEM (mm³/m³) of *P. rubescens* and *T. bourrellyi* in 2016.

T. bourrellyi started to increase in March and April, whereas a greater abundance of *P. rubescens* was detectable from the first months of the year, i.e., January and February. A temporal shift in the development of the two species can be highlighted. In January and February, the *P. rubescens* abundance had average values of 8.28 ± 2.27 mm³/m³ and 22.7 ± 6.35 mm³/m³, respectively; whereas *T. bourrellyi* recorded values of 5.6 ± 2.09 mm³/m³ and 2.78 ± 2.33 mm³/m³. An increase in *T. bourrellyi* biovolume was detected in March, in which a value of 21.5 ± 5.8 mm³/m³ was recorded, but the greatest values were detected during the next months, especially in April and May (134 ± 74 mm³/m³). The largest *P. rubescens* biovolumes were measured during March (93.9 ± 29.4 mm³/m³), and April (585 ± 21 mm³/m³). A greater biovolume of *P. rubescens* than that of *T. bourrellyi* was recorded for six months during the observation period: January, February, March, July, October, and December.

The variation of the biovolume of the two Cyanobacteria in the different months was strongly influenced both by cell density and cell volume. The species did not show differences in cell dimension at the different depths along the water column, but seasonal variations were observed. In April, *P. rubescens* considerably increased in biovolume compared to *T. bourrellyi*, as its cells displayed the highest diameter (6.33 μm), with a cell volume of 150 μm³. The average annual cell volume of *P. rubescens* and *T. bourrellyi* had the same value (63 ± 4 and 63 ± 2 μm³ respectively), with a mean annual diameter of 4.38 ± 0.11 and 4.2 ± 0.08 μm, and a mean annual length of 3.95 ± 0.04 and 4.50 ± 0.08 μm, respectively. For *P. rubescens*, the maximum cell volume was measured in April, whereas the smallest value was recorded in December (25 μm³). *T. bourrellyi* showed a maximum cell volume in October (94 μm³) and the minimum in April (38 μm³). The surface-to-volume ratio of the two species showed a similar monthly trend, with an average annual value of 1.47 and 1.43 for *P. rubescens* and *T. bourrellyi* respectively. The greatest difference between the Cyanobacteria was detected in April and January. In general terms, *P. rubescens* showed a greater annual variation in this ratio compared to *T. bourrellyi*.

The two species were widespread in the entire sampling zone, and their spatial distribution varied among the different months, with considerable concentrations in the sub-superficial layers. In considering the months with major abundance, *P. rubescens* showed the greatest biovolume around a depth of 20 m, with a relevant, although lower, presence at 10 and 30 m. *T. bourrellyi* displayed a considerable biovolume at depths of 10 and 20 m. Beyond the 40 m depth, *T. bourrellyi* abundance throughout the year was poor, whereas *P. rubescens* was present in considerable amounts at these depths, especially in the months of greater development (March, April, May) (Figure 5).

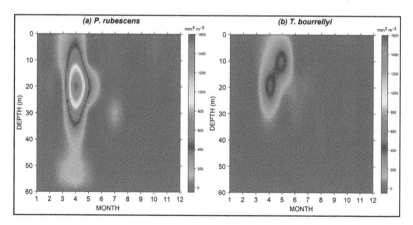

Figure 5. Contour plots showing the spatial and temporal distribution in Lake Iseo during 2016 of (**a**) *P. rubescens* biovolume (mm³/m³); (**b**) *T. bourrellyi* biovolume (mm³/m³).

3.4. Ecological Requirements of P. rubescens and T. bourrellyi

To characterize the ecological requirements of *P. rubescens* and *T. bourrellyi*, we performed principal component analysis (PCA), considering the 20 chemico-physical parameters (listed previously) and the biovolume of *P. rubescens* and *T. bourrellyi*. Below, we report only the most meaningful analysis, which includes the preliminary principal component analysis and the final analysis with properly selected observations.

The preliminary PCA showed that the eigenvalues of the first two principal components (PCs) represented 63.4% of the total variance of the observations (PC1 47.6%; PC2 15.8%) (Figure 6a).

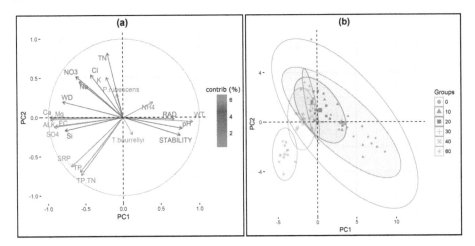

Figure 6. Preliminary principal component analysis (PCA) performed on biovolume values of *P. rubescens* and *T. bourrellyi* and 20 chemico-physical variables, detected during the whole period analyzed from the surface to a depth of 60 m. (**a**) Loading plot on the plane defined by principal components 1 (PC1) and 2 (PC2), with different colors according to the individuals' contributions ("contrib %"). See "Materials and Methods" for the abbreviations; (**b**) Score plot on the plane defined by PC1 and PC2; samples are grouped ("Groups") by different depths in ellipses (confidence level 0.95).

The first factor had the highest positive loadings for water temperature, pH, stability, and radiation; whereas the highest negative loadings were found for calcium, alkalinity, magnesium, sulphates, electrical conductivity, water density, and silicates. The second factor had the highest positive loadings for total nitrogen, and the lowest for chloride, potassium, nitrate nitrogen, and sodium; whereas negative loadings were observed for total phosphorus, and the TP:TN ratio. *P. rubescens* and *T. bourrellyi* biovolumes were poorly explained, and they were not correlated in the plane defined by PC1 and PC2. A good correlation was found between *P. rubescens* and total nitrogen; whereas *T. bourrellyi* showed a slight correlation with stability, water temperature, and pH. The score plot showed a clear difference between the observations at various depths (Figure 6b). The samples collected at a depth of 60 m displayed a different pattern, forming a cluster. This different behavior could be explained by this deep layer being the high hypolimnion, in which different conditions were able to be detected from in the superficial layers. The samples collected in April at 0 m were outliers.

Preliminary PCA allowed the identification of variables and observations on which our attention should be focused in order to determine the relationship between Cyanobacteria and environmental features. The final PCA was performed on observations from five sampling depths (excluding data collected at 60 m) over four months (February–May), in which the species started their development and showed the greatest biovolume. The eigenvalues of the first two principal components of the final PCA represented the 62.6% of the total variance, with a value of 42.0% for the first principal component, and 20.6% for the second (Figure 7).

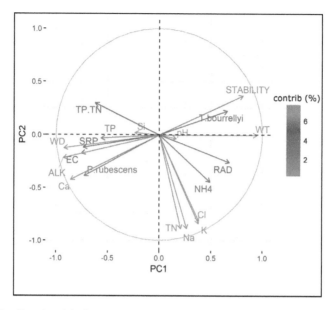

Figure 7. Loading plot of the first two components of the final principal component analysis (PCA) performed on biovolume values of *P. rubescens* and *T. bourrellyi* and 20 chemico-physical variables, detected at depths from 0 to 40 m during February, March, April, and May.

The first component had the highest positive value for water temperature, stability, radiation, and *T. bourrellyi* biovolume; and had the highest negative value for alkalinity, water density, calcium, electrical conductivity, soluble reactive phosphorus, and *P. rubescens* biovolume. Total nitrogen, sodium, potassium, and chloride were explained by the second factor, with negative loading values. A high correlation was highlighted between *T. bourrellyi* and stability, which was linked to water temperature, strengthening the relationship shown above. A correlation was also seen between this species and radiation. *P. rubescens* was highly correlated with calcium, alkalinity, electrical conductivity, and in general with the phosphorus forms.

A linear regression between *T. bourrellyi* and stability was performed to highlight the possible cause-effect relationship. We selected the months in which the species was more abundant, as previously chosen for PCA (i.e., February, March, April, May). The stability significantly and positively affected *T. bourrellyi* biovolume (F = 12.4, $p < 0.002$, $R^2 = 0.36$).

4. Discussion

A proliferation of Cyanobacteria has been recorded in lacustrine phytoplanktonic communities, and studies have highlighted the possible role of climate change in promoting toxic blooms, although considerably different responses to temperature increase have been recorded among various Cyanobacteria species and among genetic strains [8,9,12,55–60]. Moreover, these changes in lake ecosystems can lead to the introduction of allochthonous populations of Cyanobacteria, with different and not-always-identified toxicological features [4,55]. Although these complex events cannot be explained by a single environmental driver [55,61], more effort is needed to further the knowledge about the abiotic and biotic elements that promote the development of harmful Cyanobacteria.

The phytoplanktonic communities in the deep south-Alpine lakes in Italy have been repeatedly monitored and studied. A long-term data set (from 1998 to 2016) was available for Lake Iseo [21,22]. Starting in 2014, the presence of the allochthonous *T. bourrellyi* (Cyanobacteria) was detected in Lake Iseo, as well as in other south-Alpine lakes. The causes of its proliferation remain as yet poorly understood [3,4].

Our study found high biovolume values of *T. bourrellyi* in Lake Iseo, with a maximum abundance of 162 mm^3/m^3. These recorded values were comparable to those of *P. rubescens*, which was reported in previous studies as the most abundant Cyanobacteria taxa in Lake Iseo [21,22]. These two species (order Oscillatoriales) are multicellular filamentous forms, cold-stenotherm species, with a comparable growth temperature range of 10 to 20 °C [62]. A remarkable difference between the two species is related to the gas vesicles. Unlike *T. bourrellyi*, *P. rubescens* has these structures and, therefore, has an important competitive advantage. *P. rubescens* is able to move along the water column and consequently acquire the most favorable position [4,63–67]. The physiological features and environmental factors that allow the co-presence of these two Cyanobacteria have not yet been identified. In particular, understanding which environmental changes have occurred, and on what timescale, is important, as these changes did not sufficiently favor *P. rubescens* for *T. bourrellyi* to be excluded, thus permitting their co-existence [68,69].

In 2015, a stronger development of *T. bourrellyi* compared to *P. rubescens* was recorded in Lake Garda. The re-oligotrophication process of this lake has been hypothesized to be a possible explanation for the altered cyanobacterial community [4]. In the last decades of the 20th century, a decline in the *P. rubescens* population was recorded in many peri-alpine lakes, which was linked to the decrease in nutrients loading following large restoration projects, e.g., Lake Garda, Lake Maggiore [31], Lake Bourget [70], and Lake Ammersee [71]. This situation was not the case for Lake Iseo, in which nutrients have been high since the 1970s [21,22]. Moreover, for the different lakes, phytoplankton composition is largely dependent on their having similar geographic, morphometric, and hydrodynamic characteristics, whereas the biomass level is mainly determined by nutrient load [23]. The role of temperature and nutrients in affecting *P. rubescens* development is still open to considerable debate [20,72,73]: in oligotrophic lakes, the nutrients appear to be more important than the temperature; in mesotrophic lakes the temperature has a stronger impact; and in eutrophic and hyper-eutrophic lakes, the positive interaction between temperature and nutrients plays the main role [74].

On a global scale, an increasing trend in the lacustrine water temperature has been widely reported, with a mean rise of 0.34 °C/decade in the summer surface water temperature [75–77]. In particular, studies on Lake Iseo highlighted an average annual increase of 0.19 °C/decade, with a temperature rise of 0.56 °C in the 30 years from 1986 to 2015 [77]. Rising temperatures could affect the pelagic community, by both directly and indirectly influencing the growth, the survival, the metabolism, and the reproduction of organisms [8]. One of the most important consequences of this temperature increase is stronger thermal stratification, which is caused by the temperature-dependent density difference between the epilimnion and hypolimnion [78], as well as the decrease in the intensity of deep mixing, which has been observed in Lake Iseo since the mid-1980s [25]. The most recent episodes of full circulation occurred at the end of winter in 2005 and 2006, and were due to extremely cold winters and sufficient kinetic energy [35,36,79]. From 2007 until 2014, the mixing depths likely did not exceed 100 m, and the most recent years (2014–2016) have been characterized by a mixing depth accounting for no more than 20% of the water column [35,36,80]. By analyzing the available discrete data for 2016, we detected a maximum deep water mixing of 50 m.

The presence of *P. rubescens* in Lake Iseo has been reported since at least the 1980s [21,22]. The presence of gas vesicles in this species could allow the overcoming of the tendency of Cyanobacteria to sink out of the euphotic zone, providing *P. rubescens* with a competitive advantage. Changes in cellular weight due to photosynthesis and respiration processes can be balanced with gas vesicles [65,66,81]. Moreover, the buoyancy ability allows each species to optimize light absorption and obtain the most favorable positioning in the water column. This species, being tolerant to light deficiency, tends to form a layer on the lower border of the metalimnion in deep and stratified lakes, where it can avoid interspecific competition and exploit resources and nutrients [65,74,82]. The gas vesicles can, however, irreversibly collapse at a certain critical hydrostatic pressure and then no longer provide buoyancy. Consequently, the depth to which the planktonic organisms can be mixed in winter is of paramount importance, determining the amount of Cyanobacteria that remain vital at the end of the winter and that form the inoculum for growth during the next season [23,83]. D'Alelio et al. [84] demonstrated that the presence of weaker or stronger gas vesicles in *P. rubescens* is an adaptive based on shallow or deep winter circulation, respectively. In 2009 in Lake Iseo they also underlined the presence of *P. rubescens* strains with strong gas vesicles and a critical collapse pressure (p_c) of $\cong 1.1$ MPa that were able to overcome hydrostatic pressure of water depths of up to 80–100 m in winter. Given the reduced mixing depth recorded in Lake Iseo in recent decades, *P. rubescens* was able to maintain a reproductive population during the winter months with weak vertical circulation, i.e., January and February. For this reason, and given the ability of *P. rubescens* to use limited solar radiation, this population was able to grow early in the first months of the year, gaining a competitive advantage over other phytoplanktonic taxa, e.g., *T. bourrellyi*.

T. bourrellyi does not possess gas vesicles, and, during calm, thermally stratified conditions, sinks to form a metalimnetic maximum [67]. As this species does not have a buoyancy ability, a different strategy is used to optimize light absorption. These features make this species well-adapted to stratified lakes [67]. A deep-water mixing would bring this organism out of the euphotic zone, where the conditions are not suitable for its survival. Our PCA analyses are in agreement with this statement, showing a strong relationship between *T. bourrellyi* and the parameters of stability (Brunt-Väisälä frequency) and radiation. Then, when the stratification was more obvious and the radiation higher, the species began its development. The nutrients, instead, seemed to be a less critical factor in promoting one species over the other.

Despite the shift in spring development, both Cyanobacteria attained maximum biovolume in April; the biovolume then decreased from June to January. The same temporal development of *T. bourrellyi* was highlighted in Lake Garda (deep south-Alpine lake) in 2015, with the maximum reached during April and May, and a remarkable decrease taking place in August [4], confirming that the most suitable conditions for this species occurred during the spring months. The extremely high biovolume of *P. rubescens* in April was related to the elevated cell volume, especially cell width, as the densities of the two Cyanobacteria were comparable. At present, clearly evaluating what adaptive significance the variation in cellular dimensions could have is not possible. We only speculated that the greater cellular dimension observed could be related to the increase in gas vesicle diameters. In April, a more efficient buoyancy capacity, proportional to vesicle volume [85], could favor the movement of the organism in the stable water column. In our work, gas vesicle diameter measurement was not performed. In the literature, to the best of our knowledge, no one has verified this hypothesis. However, the inverse relationship between the gas vesicle width and the critical pressure at which they collapse has been documented [85,86]. Therefore, *P. rubescens*, during the mixing period in which the strains could be carried at depths with elevated hydrostatic pressure, developed smaller and stronger gas vesicles, to avoid their collapse. The enhanced stability condition allowed the increase of the gas vesicles diameter and consequently the movement of the organisms became more efficient. The increase in cellular dimension involved a change in the surface to volume ratio, influencing nutrients and light absorption [87]. With the increased cell width, the ratio became unfavorable in April for *P. rubescens*.

P. rubescens attained a maximum biovolume around a depth of 20 m; *T. bourrellyi* showed a different pattern with a spatial shift from April (20 m) to May (10 m). This behavior could be linked to the upward repositioning of the thermocline, detected from April ($z_{therm} = 15.7$ m) to May ($z_{therm} = 13.1$ m) (Figures 2 and 5). The vertical distribution of *T. bourrellyi* was tied to the thermocline positioning in the water column, as this species does not possess gas vesicles and cannot independently move in the water column, and accumulated in alignment with the major density-temperature discontinuity [4,67]. The presence of gas vesicles in *P. rubescens* allowed this species to move along the water column and grow in the layer with the best irradiance conditions, which were detected in Lake Iseo at a depth of around 20 m. This species is also well adapted to low light values, and has been reported to be an efficient light harvester [65,81,88]. It was detected growing in the radiation conditions between 40 and 60 m, where it developed with lesser biovolume values than in the upper layers.

In conclusion, this work assessed the environmental changes that have occurred in Lake Iseo that permitted the settlement of the allochthonous Cyanobacteria *T. bourrellyi*, and the co-existence with *P. rubescens*, which still had a large biovolume. Our results confirmed that *P. rubescens* was competitive not only in stratified lakes, but also in lacustrine ecosystems characterized by partial mixing in the water column. Its survival was due to its features; *P. rubescens* was able to survive the winter, and started to grow immediately when allowed by the radiation and stability conditions. The increased stratified conditions in the aquatic system seemed to be a key factor for the settling of *T. bourrellyi*, which cannot proliferate in lakes with deep mixing events. These results suggest the possible role of enhanced stability in promoting the development of this allochthonous Cyanobacteria, which has created its own niche as a result of the more stable water column and reduced mixing depths in the winter. Climate changes can directly and indirectly affect aquatic environments, and more effort is needed to better understand the modified dynamics in lakes, and to expand the knowledge obtained by annual investigation, to understand if what was verified in Lake Iseo could be found in the other southern Alpine lakes.

Acknowledgments: We would like to dedicate our work to the dear colleague Giuseppe Morabito, whose presence and competences we miss. This work was supported by Fondazione Cariplo, grant 2014-1282, and by the University of Milano-Bicocca (FA). We are grateful to ARPA Brescia, to the Provincial Police of Brescia and to the Police Officer Cesare Poli for logistic support in the field. We especially want to thank Nico Salmaso for knowledge-transfer support, Pietro Fumagalli for help during the field sampling, and Morena Spreafico for her valuable review contribution. We thank the two anonymous reviewers for their comments, which helped to improve this article.

Author Contributions: V.N. analyzed the phytoplankton analyses, ran the statistical analyses, elaborated the data and wrote the manuscript; M.P. planned and made the field measurements, collected samples and analyzed phytoplankton samples; V.S. made the chemical analyses; B.L. conceived the study, made the field measurements, revised and improved the manuscript.

Conflicts of Interest: The authors declare no conflict of interest.

References

1. Paerl, H.W. Controlling cyanobacterial harmful blooms in freshwater ecosystems. *Microb. Biotechnol.* **2017**, *10*, 1106–1110. [CrossRef] [PubMed]
2. Bekker, A.; Holland, H.D.; Wang, P.L.; Rumble, D.; Stein, H.J.; Hannah, J.L.; Coetzee, L.L.; Beukes, N.J. Dating the rise of atmospheric oxygen. *Nature* **2004**, *427*, 117–120. [CrossRef] [PubMed]
3. Shams, S.; Capelli, C.; Cerasino, L.; Ballot, A.; Dietrich, D.R.; Sivonen, K.; Salmaso, N. Anatoxin-a producing *Tychonema* (Cyanobacteria) in European waterbodies. *Water Res.* **2015**, *69*, 68–79. [CrossRef] [PubMed]
4. Salmaso, N.; Cerasino, L.; Boscaini, A.; Capelli, C. Planktic *Tychonema* (Cyanobacteria) in the large lakes south of the alps: Phylogenetic assessment and toxigenic potential. *FEMS Microbiol. Ecol.* **2016**, *92*, fiw155. [CrossRef] [PubMed]
5. Paerl, H.W.; Otten, T.G. Harmful Cyanobacterial Blooms: Causes, Consequences, and Controls. *Microb. Ecol.* **2013**, *65*, 995–1010. [CrossRef] [PubMed]
6. Jeppesen, E.; Kronvang, B.; Meerhoff, M.; Søndergaard, M.; Hansen, K.M.; Andersen, H.E.; Lauridsen, T.L.; Liboriussen, L.; Beklioglu, M.; Özen, A.; et al. Climate Change Effects on Runoff, Catchment Phosphorus Loading and Lake Ecological State, and Potential Adaptations. *J. Environ. Qual.* **2009**, *38*, 1930–1941. [CrossRef] [PubMed]
7. Wagner, C.; Adrian, R. Cyanobacteria dominance: Quantifying the effects of climate change. *Limnol. Oceanogr.* **2009**, *54*, 2460–2468. [CrossRef]

8.   Carey, C.C.; Ibelings, B.W.; Hoffmann, E.P.; Hamilton, D.P.; Brookes, J.D. Eco-physiological adaptations that favour freshwater cyanobacteria in a changing climate. *Water Res.* **2012**, *46*, 1394–1407. [CrossRef] [PubMed]

9.   Paerl, H.W.; Paul, V.J. Climate change: Links to global expansion of harmful cyanobacteria. *Water Res.* **2012**, *46*, 1349–1363. [CrossRef] [PubMed]

10. Zhang, M.; Duan, H.; Shi, X.; Yu, Y.; Kong, F. Contributions of meteorology to the phenology of cyanobacterial blooms: Implications for future climate change. *Water Res.* **2012**, *46*, 442–452. [CrossRef] [PubMed]

11. Sukenik, A.; Quesada, A.; Salmaso, N. Global expansion of toxic and non-toxic cyanobacteria: Effect on ecosystem functioning. *Biodivers. Conserv.* **2015**, *24*, 889–908. [CrossRef]

12. Paerl, H.W.; Gardner, W.S.; Havens, K.E.; Joyner, A.R.; McCarthy, M.J.; Newell, S.E.; Qin, B.; Scott, J.T. Mitigating cyanobacterial harmful algal blooms in aquatic ecosystems impacted by climate change and anthropogenic nutrients. *Harmful Algae* **2016**, *54*, 213–222. [CrossRef] [PubMed]

13. Visser, P.M.; Verspagen, J.M.H.; Sandrini, G.; Stal, L.J.; Matthijs, H.C.P.; Davis, T.W.; Paerl, H.W.; Huisman, J. How rising CO_2 and global warming may stimulate harmful cyanobacterial blooms. *Harmful Algae* **2016**, *54*, 145–159. [CrossRef] [PubMed]

14. Reichwaldt, E.S.; Ghadouani, A. Effects of rainfall patterns on toxic cyanobacterial blooms in a changing climate: Between simplistic scenarios and complex dynamics. *Water Res.* **2012**, *46*, 1372–1393. [CrossRef] [PubMed]

15. Paerl, H.W.; Joyner, J.J.; Joyner, A.R.; Arthur, K.; Paul, V.; O'Neil, J.M.; Heil, C.A. Co-occurrence of dinoflagellate and cyanobacterial harmful algal blooms in southwest Florida coastal waters: Dual nutrient (N and P) input controls. *Mar. Ecol. Prog. Ser.* **2008**, *371*, 143–153. [CrossRef]

16. Sukenik, A.; Hadas, O.; Kaplan, A.; Quesada, A. Invasion of Nostocales (cyanobacteria) to subtropical and temperate freshwater lakes—Physiological, regional, and global driving forces. *Front. Microbiol.* **2012**, *3*, 86. [CrossRef] [PubMed]

17. Carmichael, W.W.; Boyer, G.L. Health impacts from cyanobacteria harmful algae blooms: Implications for the North American Great Lakes. *Harmful Algae* **2016**, *54*, 194–212. [CrossRef] [PubMed]

18. Meriluoto, J.; Blaha, L.; Bojadzija, G.; Bormans, M.; Brient, L.; Codd, G.A.; Drobac, D.; Faassen, E.J.; Fastner, J.; Hiskia, A.; et al. Toxic cyanobacteria and cyanotoxins in European waters—Recent progress achieved through the CYANOCOST Action and challenges for further research. *Adv. Oceanogr. Limnol.* **2017**, *8*, 161–178. [CrossRef]

19. Manganelli, M.; Stefanelli, M.; Vichi, S.; Andreani, P.; Nascetti, G.; Scialanca, F.; Scardala, S.; Testai, E.; Funari, E. Cyanobacteria biennal dynamic in a volcanic mesotrophic lake in central Italy: Strategies to prevent dangerous human exposures to cyanotoxins. *Toxicon* **2016**, *115*, 28–40. [CrossRef] [PubMed]

20. Gallina, N.; Beniston, M.; Jacquet, S. Estimating future cyanobacterial occurrence and importance in lakes: A case study with *Planktothrix rubescens* in Lake Geneva. *Aquat. Sci.* **2017**, *79*, 249–263. [CrossRef]

21. Garibaldi, L.; Anzani, A.; Marieni, A.; Leoni, B.; Mosello, R. Studies on the phytoplankton of the deep subalpine Lake Iseo. *J. Limnol.* **2003**, *62*, 177–189. [CrossRef]

22. Leoni, B.; Marti, C.L.; Imberger, J.; Garibaldi, L. Summer spatial variations in phytoplankton composition and biomass in surface waters of a warm-temperate, deep, oligo-holomictic lake: Lake Iseo, Italy. *Inland Waters* **2014**, *4*, 303–310. [CrossRef]

23. Morabito, G.; Ruggiu, D.; Panzani, P. Recent dynamics (1995–1999) of the phytoplankton assemblages in Lago Maggiore as a basic tool for defining association patterns in the Italian deep lakes. *J. Limnol.* **2002**, *61*, 129–145. [CrossRef]

24. Salmaso, N. Seasonal variation in the composition and rate of change of the phytoplankton community in a deep subalpine lake (Lake Garda, Northern Italy). An application of nonmetric multidimensional scaling and cluster analysis. *Hydrobiologia* **1996**, *337*, 49–68. [CrossRef]

25. Salmaso, N.; Mosello, R. Limnological research in the deep southern subalpine lakes: Synthesis, directions and perspectives. *Adv. Oceanogr. Limnol.* **2010**, *1*, 29–66. [CrossRef]

26. Ruggiu, D.; Morabito, G.; Panzani, P.; Pugnetti, A. Trends and relations among basic phytoplankton characteristics in the course of the long-term oligotrophication of Lake Maggiore (Italy). *Hydrobiologia* **1998**, *369*, 243–257. [CrossRef]

27. Ambrosetti, W.; Barbanti, L.; Mosello, R.; Pugnetti, A. Limnological studies on the deep southern Alpine lakes Maggiore, Lugano, Como, Iseo and Garda. *Mem. Ist. Ital. Idrobiol.* **1992**, *50*, 117–146.

28. Manca, M.; Bertoni, R. Seventy five years of limnology at the Istituto Italiano di Idrobiologia in Pallanza. *J. Limnol.* **2014**, *73*, 5–19. [CrossRef]

29. Anagnostidis, K.; Komárek, J. Modern approach to the classification system of cyanophytes. 3. Oscillatoriales. *Arch. Hydrobiol.* **1988**, *80*, 327–472.

30. Shao, J.; Peng, L.; Luo, S.; Yu, G.; Gu, J.; Lin, S.; Li, R. First report on the allelopathic effect of *Tychonema bourrellyi* (Cyanobacteria) against *Microcystis aeruginosa* (Cyanobacteria). *J. Appl. Phycol.* **2013**, *25*, 1567–1573. [CrossRef]

31. Salmaso, N.; Morabito, G.; Mosello, R.; Garibaldi, L.; Simona, M.; Buzzi, F.; Ruggiu, D. A synoptic study of phytoplankton in the deep lakes south of the Alps (lakes Garda, Iseo, Como, Lugano and Maggiore). *J. Limnol.* **2003**, *62*, 207–227. [CrossRef]

32. Winder, M.; Sommer, U. Phytoplankton response to a changing climate. *Hydrobiologia* **2012**, *698*, 5–16. [CrossRef]

33. Dokulil, M.T. Impact of climate warming on European inland waters. *Inl. Waters* **2014**, *4*, 27–40. [CrossRef]

34. Mosello, R.; Ambrosetti, W.; Arisci, S.; Bettinetti, R.; Buzzi, F.; Calderoni, A.; Carrara, E.; De Bernardi, R.; Galassi, S.; Garibaldi, L.; et al. Evoluzione recente della qualità delle acque dei laghi profondi sudalpini (Maggiore, Lugano, Como, Iseo e Garda) in risposta alle pressioni antropiche e alle variazioni climatiche. *Biol. Ambient.* **2010**, *24*, 167–177.

35. Rogora, M.; Buzzi, F.; Dresti, C.; Leoni, B.; Lepori, F.; Mosello, R.; Patelli, M.; Salmaso, N. Climatic effects on vertical mixing and deep-water oxygenation in the deep subalpine lakes in Italy. *Hydrobiologia* **2017**, accepted for publication.

36. Leoni, B.; Garibaldi, L.; Gulati, R.D. How does interannual trophic variability caused by vertical water mixing affect reproduction and population density of the *Daphnia longispina* group in Lake Iseo, a deep stratified lake in Italy? *Inland Waters* **2014**, *4*, 193–203. [CrossRef]

37. Marti, C.M.; Imberger, J.; Garibaldi, L.; Leoni, B. Using time scales to characterize phytoplankton assemblages in a deep subalpine lake during the thermal stratification period: Lake Iseo, Italy. *Water Resour. Res.* **2015**, *52*, 1762–1780. [CrossRef]

38. Pilotti, M.; Valerio, G.; Leoni, B. Data set for hydrodynamic lake model calibration: A deep prealpine case. *Water Resour. Res.* **2013**, *49*, 7159–7163. [CrossRef]

39. Valerio, G.; Pilotti, M.; Barontini, S.; Leoni, B. Sensitivity of the multiannual thermal dynamics of a deep pre-alpine lake to climatic change. *Hydrol. Process.* **2015**, *29*, 767–779. [CrossRef]

40. Leoni, B. Zooplankton predators and preys: Body size and stable isotope to investigate the pelagic food web in a deep lake (Lake Iseo, Northern Italy). *J. Limnol.* **2017**, *76*, 85–93. [CrossRef]

41. Graham, M.H.; Mitchell, B.G. Obtaining absorption spectra from individual macroalgal spores using microphotometry. *Hydrobiologia* **1999**, *398/399*, 231–239. [CrossRef]

42. Komárek, J.; Albertano, P. Cell structure of a planktic cyanoprokaryote, *Tychonema bourrellyi. Algol. Stud.* **1994**, *75*, 157–166.

43. Hillebrand, H.; Dürselen, C.-D.; Kirschtel, D.; Pollingher, U.; Zohary, T. Biovolume Calculation for Pelagic and Benthic Microalgae. *J. Phycol.* **1999**, *35*, 403–424. [CrossRef]

44. APHA-AWWA-WEF. *Standard Methods for the Examination of Water and Wastewater*, 20th ed.; American Public Health Association: Washington, DC, USA, 1998; ISBN 0875532357.

45. Salmaso, N. Factors affecting the seasonality and distribution of cyanobacteria and chlorophytes: A case study from the large lakes south of the Alps, with special reference to Lake Garda. *Hydrobiologia* **2000**, *438*, 43–63. [CrossRef]

46. Winslow, L.; Read, J.; Woolway, R.; Brentrup, J.; Leach, T.; Zwart, J. rLakeAnalyzer: Lake Physics Tools. 2017. Available online: https://cran.r-project.org/web/packages/rLakeAnalyzer/rLakeAnalyzer.pdf (accessed on 17 December 2017).

47. Imberger, J.; Patterson, J.C. Physical Limnology. *Adv. Appl. Mech.* **1989**, *27*, 303–475. [CrossRef]

48. Jollife, I.T.; Cadima, J. Principal component analysis: A review and recent developments. *Philos. Trans. R. Soc. A Math. Phys. Eng. Sci.* **2016**, *374*, 20150202. [CrossRef] [PubMed]

49. Zuur, A.F.; Ieno, E.N.; Smith, G.M. *Analysing Ecological Data*, 1st ed.; Springer: New York, NY, USA, 2007; ISBN 0-387-45967-7.

50. Shapiro, S.S.; Wilk, M.B. An Analysis of Variance Test for Normality (Complete Samples). *Biometrika* **1965**, *52*, 591–611. [CrossRef]

51. Borcard, D.; Gillet, F.; Legendre; Legendre, P. *Numerical Ecology with R*, 1st ed.; Springer: New York, NY, USA, 2011; ISBN 9788578110796.

52. Legendre, P.; Gallagher, E.D. Ecologically meaningful transformations for ordination of species data. *Oecologia* **2001**, *129*, 271–280. [CrossRef] [PubMed]

53. R Core Team. *R: A Language and Environment for Statistical Computing*; R Core Team: Vienna, Austria, 2017; ISBN 3_900051_00_3.

54. Kraemer, B.M.; Anneville, O.; Chandra, S.; Dix, M.; Kuusisto, E.; Livingstone, D.M.; Rimmer, A.; Schladow, S.G.; Silow, E.; Sitoki, L.M.; et al. Morphometry and average temperature affect lake stratification responses to climate change. *Geophys. Res. Lett.* **2015**, *42*, 4981–4988. [CrossRef]

55. O'Neil, J.M.; Davis, T.W.; Burford, M.A.; Gobler, C.J. The rise of harmful cyanobacteria blooms: The potential roles of eutrophication and climate change. *Harmful Algae* **2012**, *14*, 313–334. [CrossRef]

56. Zhang, M.; Qin, B.; Yu, Y.; Yang, Z.; Shi, X.; Kong, F. Effects of temperature fluctuation on the development of cyanobacterial dominance in spring: Implication of future climate change. *Hydrobiologia* **2016**, *763*, 135–146. [CrossRef]

57. Wood, S.A.; Borges, H.; Puddick, J.; Biessy, L.; Atalah, J.; Hawes, I.; Dietrich, D.R.; Hamilton, D.P. Contrasting cyanobacterial communities and microcystin concentrations in summers with extreme weather events: Insights into potential effects of climate change. *Hydrobiologia* **2017**, *785*, 71–89. [CrossRef]

58. Gallina, N.; Anneville, O.; Beniston, M. Impacts of extreme air temperatures on cyanobacteria in five deep peri-alpine lakes. *J. Limnol.* **2011**, *70*, 186–196. [CrossRef]

59. Havens, K.; Paerl, H.; Phlips, E.; Zhu, M.; Beaver, J.; Srifa, A. Extreme weather events and climate variability provide a lens to how shallow lakes may respond to climate change. *Water* **2016**, *8*, 229. [CrossRef]

60. Lürling, M.; Eshetu, F.; Faassen, E.J.; Kosten, S.; Huszar, V.L.M. Comparison of cyanobacterial and green algal growth rates at different temperatures. *Freshw. Biol.* **2013**, *58*, 552–559. [CrossRef]

61. Heisler, J.; Glibert, P.M.; Burkholder, J.M.; Anderson, D.M.; Cochlan, W.; Dennison, W.C.; Dortch, Q.; Gobler, C.J.; Heil, C.A.; Humphries, E.; et al. Eutrophication and harmful algal blooms: A scientific consensus. *Harmful Algae* **2008**, *8*, 3–13. [CrossRef] [PubMed]

62. Suda, S.; Watanabe, M.M.; Otsuka, S.; Mahakahant, A.; Yongmanitchai, W.; Nopartnaraporn, N.; Liu, Y.; Day, J.G. Taxonomic revision of water-bloom-forming species of oscillatorioid cyanobacteria. *Int. J. Syst. Evol. Microbiol.* **2002**, *52*, 1577–1595. [CrossRef] [PubMed]

63. Bright, D.I.; Walsby, A.E. The relationship between critical pressure and width of gas vesicles in isolates of *Planktothrix rubescens* from Lake Zurich. *Microbiology* **1999**, *145*, 2769–2775. [CrossRef] [PubMed]

64. Garneau, M.È.; Posch, T.; Pernthaler, J. Seasonal patterns of microcystin-producing and non-producing *Planktothrix rubescens* genotypes in a deep pre-alpine lake. *Harmful Algae* **2015**, *50*, 21–31. [CrossRef]

65. Kurmayer, R.; Deng, L.; Entfellner, E. Role of toxic and bioactive secondary metabolites in colonization and bloom formation by filamentous cyanobacteria *Planktothrix*. *Harmful Algae* **2016**, *54*, 69–86. [CrossRef] [PubMed]

66. Pancrace, C.; Barny, M.-A.; Ueoka, R.; Calteau, A.; Scalvenzi, T.; Pédron, J.; Barbe, V.; Piel, J.; Humbert, J.-F.; Gugger, M. Insights into the *Planktothrix* genus: Genomic and metabolic comparison of benthic and planktic strains. *Sci. Rep.* **2017**, *7*, 41181. [CrossRef] [PubMed]

67. Ganf, G.G.; Heaney, S.I.; Corry, J. Light-Absorption and Pigment Content in Natural-Populations and Cultures of a Non-Gas Vacuolate Cyanobacterium *Oscillatoria bourrellyi* (=*Tychonema bourrellyi*). *J. Plankton Res.* **1991**, *13*, 1101–1121. [CrossRef]

68. Li, L.; Chesson, P. The Effects of Dynamical Rates on Species Coexistence in a Variable Environment: The Paradox of the Plankton Revisited. *Am. Nat.* **2016**, *188*, E46–E58. [CrossRef] [PubMed]

69. Hutchinson, G.E. The Paradox of the Plankton. *Am. Nat.* **1961**, *95*, 137–145. [CrossRef]

70. Jacquet, S.; Briand, J.F.; Leboulanger, C.; Avois-Jacquet, C.; Oberhaus, L.; Tassin, B.; Vinçon-Leite, B.; Paolini, G.; Druart, J.C.; Anneville, O.; et al. The proliferation of the toxic cyanobacterium *Planktothrix rubescens* following restoration of the largest natural French lake (Lac du Bourget). *Harmful Algae* **2005**, *4*, 651–672. [CrossRef]

71. Ernst, B.; Hoeger, S.J.; O'Brien, E.; Dietrich, D.R. Abundance and toxicity of *Planktothrix rubescens* in the pre-alpine Lake Ammersee, Germany. *Harmful Algae* **2009**, *8*, 329–342. [CrossRef]

72. Dokulil, M.T.; Teubner, K. Deep living *Planktothrix rubescens* modulated by environmental constraints and climate forcing. *Hydrobiologia* **2012**, *698*, 29–46. [CrossRef]

73. Savichtcheva, O.; Debroas, D.; Perga, M.E.; Arnaud, F.; Villar, C.; Lyautey, E.; Kirkham, A.; Chardon, C.; Alric, B.; Domaizon, I. Effects of nutrients and warming on *Planktothrix* dynamics and diversity: A palaeolimnological view based on sedimentary DNA and RNA. *Freshw. Biol.* **2015**, *60*, 31–49. [CrossRef]

74. Rigosi, A.; Carey, C.C.; Ibelings, B.W.; Brookes, J.D. The interaction between climate warming and eutrophication to promote cyanobacteria is dependent on trophic state and varies among taxa. *Limnol. Oceanogr.* **2014**, *59*, 99–114. [CrossRef]

75. Salmaso, N.; Buzzi, L.; Cerasino, L.; Garibaldi, L.; Leoni, B.; Manca, M.; Morabito, G.; Rogora, M.; Simona, M. Influenza delle fluttuazioni climatiche sui grandi laghi a sud delle Alpi: Implicazioni nel contesto del riscaldamento globale. *Biol. Ambient.* **2014**, *28*, 17–32.

76. O'Reilly, C.M.; Rowley, R.J.; Schneider, P.; Lenters, J.D.; Mcintyre, P.B.; Kraemer, B.M. Rapid and highly variable warming of lake surface waters around the globe. *Geophys. Res. Lett.* **2015**, *42*, 10773–10781. [CrossRef]

77. Pareeth, S.; Bresciani, M.; Buzzi, F.; Leoni, B.; Lepori, F.; Ludovisi, A.; Morabito, G.; Adrian, R.; Neteler, M.; Salmaso, N. Warming trends of perialpine lakes from homogenised time series of historical satellite and in-situ data. *Sci. Total Environ.* **2017**, *578*, 417–426. [CrossRef] [PubMed]

78. Posch, T.; Köster, O.; Salcher, M.M.; Pernthaler, J. Harmful filamentous cyanobacteria favoured by reduced water turnover with lake warming. *Nat. Clim. Chang.* **2012**, *2*, 809–813. [CrossRef]

79. Salmaso, N.; Morabito, G.; Garibaldi, L.; Mosello, R. Trophic development of the deep lakes south of the Alps: A comparative analysis. *Arch. Hydrobiol.* **2007**, *170*, 177–196. [CrossRef]

80. Leoni, B.; Nava, V.; Patelli, M. Relationship among climate variability, Cladocera phenology and the pelagic food web in deep lakes in different trophic states. *Mar. Freshw. Res.* **2017**, accepted for publication.

81. Walsby, A.E.; Schanz, F. Light-dependent growth rate determines changes in the population of *Planktothrix rubescens* over the annual cycle in lake Zürich, Switzerland. *New Phytol.* **2002**, *154*, 671–687. [CrossRef]

82. Mantzouki, E.; Visser, P.M.; Bormans, M.; Ibelings, B.W. Understanding the key ecological traits of cyanobacteria as a basis for their management and control in changing lakes. *Aquat. Ecol.* **2016**, *50*, 333–350. [CrossRef]

83. Walsby, A.E.; Avery, A.; Schanz, F. The critical pressures of gas vesicles in *Planktothrix rubescens* in relation to the depth of winter mixing in Lake Zurich, Switzerland. *J. Plankton Res.* **1998**, *20*, 1357–1375. [CrossRef]

84. D'Alelio, D.; Gandolfi, A.; Boscaini, A.; Flaim, G.; Tolotti, M.; Salmaso, N. *Planktothrix* populations in subalpine lakes: Selection for strains with strong gas vesicles as a function of lake depth, morphometry and circulation. *Freshw. Biol.* **2011**, *56*, 1481–1493. [CrossRef]

85. Walsby, A.E.; Bleything, A. The Dimensions of Cyanobacterial Gas Vesicles in Relation to Their Efficiency in Providing Buoyancy and Withstanding Pressure. *J. Gen. Microbiol.* **1988**, *134*, 2635–2645. [CrossRef]

86. Hayes, P.K.; Walsby, A.E. The inverse correlation between width and strength of gas vesicles in cyanobacteria. *Br. Phycol. J.* **1986**, *21*, 191–197. [CrossRef]

87. Reynolds, C.S.; Alex Elliott, J.; Frassl, M.A. Predictive utility of trait-separated phytoplankton groups: A robust approach to modeling population dynamics. *J. Great Lakes Res.* **2014**, *40*, 143–150. [CrossRef]

88. Yankova, Y.; Villiger, J.; Pernthaler, J.; Schanz, F.; Posch, T. Prolongation, deepening and warming of the metalimnion change habitat conditions of the harmful filamentous cyanobacterium *Planktothrix rubescens* in a prealpine lake. *Hydrobiologia* **2016**, *776*, 125–138. [CrossRef]

Article

Effect of Temperature Rising on the Stygobitic Crustacean Species *Diacyclops belgicus*: Does Global Warming Affect Groundwater Populations?

Tiziana Di Lorenzo [1],* and Diana Maria Paola Galassi [2]

[1] Institute of Ecosystem Study of the CNR—National Research Council of Italy, Via Madonna del Piano 10, 50019 Sesto Fiorentino, Italy

[2] Department of Life, Health and Environmental Sciences, University of L'Aquila, Via Vetoio 1, 67100 Coppito, Italy; dianamariapaola.galassi@univaq.it

*   Correspondence: tiziana.dilorenzo@ise.cnr.it; Tel.: +39-055-522-5918

Received: 27 October 2017; Accepted: 4 December 2017; Published: 7 December 2017

Abstract: The average global temperature is predicted to increase by 3 °C by the end of this century due to human-induced climate change. The overall metabolism of the aquatic biota will be directly affected by rising temperatures and associated changes. Since thermal stability is a characteristic of groundwater ecosystems, global warming is expected to have a profound effect on the groundwater fauna. The prediction that stygobitic (obligate groundwater dweller) species are vulnerable to climate change includes assumptions about metabolic effects that can only be tested by comparisons across a thermal gradient. To this end, we investigated the effects of two different thermal regimes on the metabolism of the stygobitic copepod species *Diacyclops belgicus* (Kiefer, 1936). We measured the individual-based oxygen consumption of this species as a proxy of possible metabolic reactions to temperature rising from 14 to 17 °C. We used a sealed glass microplate equipped with planar oxygen sensor spots with optical isolation glued onto the bottom of 80-µL wells integrated with a 24-channel fluorescence-based respirometry system. The tests have provided controversial results according to which the *D. belgicus* populations should be prudently considered at risk under a global warming scenario.

Keywords: metabolism; respirometry; copepods; crustaceans; groundwater; porous aquifer

1. Introduction

Global climate change is expected to seriously alter the supply of aquatic ecosystem services that are crucial for human wellbeing. Mediterranean ecosystems appear the most vulnerable to global change in Europe, with potential impacts related primarily to increasing temperatures, reduction in precipitation, water scarcity, concentration of economic activities in coastal areas and climate-sensitive agriculture [1]. Groundwater will play a fundamental role in sustaining ecosystems and enabling human adaptation to climate change. The strategic importance of groundwater will intensify as climate extremes become more frequent and intense. However, understanding climate-change effects on groundwater ecosystems is not an easy task because global warming may affect groundwater ecosystems both directly, through changes in temperature and replenishment by recharge, and indirectly, through changes in groundwater use [2,3]. Accordingly, the IPPC (Intergovernmental Panel on Climate Change) [4] stated that a lack of necessary data has made it impossible to determine the magnitude and direction of groundwater change due to climate change [5]. Multidisciplinary collaboration is needed to study changes in groundwater chemistry, temperature, hydro-geophysical properties and biology over a range of spatial scales [6].

An air temperature increase of 2–3 °C is expected in the Mediterranean region by 2050 and a rise of 3–5 °C is expected by 2100 [4,7]. Global warming will result in an increase of groundwater temperatures as well. In the extreme cases, groundwater temperatures in shallow porous aquifers (depth < 15 m below ground level) are expected to increase by up to 4–5 °C in some temperate climate regions in the 45° northern latitude, depending on the degree of urbanization [8]. Climate warming will affect the aquifers that are recharged by surface water through riverbank infiltration [9] as well as those that are directly recharged by precipitation and those affected by ground surface heating [8]. In shallow porous aquifers temperatures are not constant. Seasonal temperature cycles at the ground surface drive seasonal temperature fluctuations in the subsurface down to depths of 10–15 m [10]. Diurnal temperature fluctuations are typically found at depths of less than 1 m, but seasonal fluctuations may be detectable at depths up to 15 m [10]. The annual amplitude of temperature is typically less than 2 °C in porous aquifers that are deeper than 5 m below the soil surface [11–15]. Yet, shallow porous aquifers are not immune to global warming, because the annual mean temperature of groundwater closely tracks the ambient air temperature [11]. This means that groundwater temperatures in these aquifers might reach high values in the warm seasons under the expected global warming scenarios.

While the effects of global warming on groundwater chemistry, hydro-geophysical properties and resources have been studied in recent years, the assessment of the biological responses to the groundwater temperature increases due to climate change is in its infancy. Metabolic rates of obligate groundwater (stygobitic) species are expected to be highly temperature-dependent. Groundwater ectotherm species reside in habitats where temperatures may vary seasonally but only one or two degrees throughout the year [16,17]. In such stable thermal environments, natural selection should favor stenothermal species, i.e., organisms that maximize their performance along a very narrow range of temperatures [18]. Few experiments that deal with the impact of temperature on the physiology of obligate groundwater species are available [19–23], however none of them was specifically designed to test the predictions of the climate change hypothesis.

In this study, we determined the thermal tolerance of a stygobitic species, *Diacyclops belgicus* (Kiefer, 1936), to increasing groundwater temperatures. We selected *D. belgicus* because it is an obligate groundwater species that has been collected from caves, porous aquifers and hyporheic zones of Europe and the former USSR (the Union of Soviet Socialist Republics) [24]. The species shows a wide geographic distribution, and no marked habitat and microhabitat preferences, even if it has been more frequently collected from porous aquifers and the hyporheic zone. *D. belgicus* also resides in a Mediterranean shallow porous aquifer in Tuscany which is characterized by annual temperature amplitude of about 2 °C. Since an exponential relationship between oxygen consumption and temperature was observed in some other obligate groundwater species [23], we expected a change of the metabolic rates of *D. belgicus* under rising temperatures. We explored the likelihood of this prediction by measuring the oxygen consumption of *D. belgicus* under two different temperatures, namely 14 and 17 °C. Fourteen degrees is the lowest annual temperature that has been measured in the shallow porous aquifer where the individuals of *D. belgicus* that were used in our trials were collected. The mean annual temperature of the collection site of *D. belgicus* in this study is 15 °C. Seventeen degrees is 1.1 °C above the highest annual temperature recorded (15.9 °C). The aquifer temperature was measured monthly by a multiparametric probe (ECM MultiTM; Lange GmbH, Düsseldorf, Germany). We selected 14 °C as the reference value because this temperature appeared to be an important threshold for some stygobitic crustaceans, namely the amphipods *Niphargus rhenorhodanensis* Schellenberg, 1937 and *N. virei* Chevreux, 1896: the ventilatory activity increased largely above this temperature in these organisms [20]. The temperature of 17 °C was selected because a 2 °C increase in the mean annual groundwater temperature is anticipated by 2050 due to global warming.

2. Materials and Methods

2.1. Animal Collection and Rearing

The specimens of *D. belgicus* (Figure 1) were collected from a phreatic well in Tuscany, Italy (coordinates: 43°49′02.61″ N; 11°11′59.79″ E) in June 2017. The well (depth: 14 m) is situated in the shallow Quaternary porous aquifer of Medio Valdarno. Prior to the collection of the animals, some groundwater samples were withdrawn from the well and analyzed for dissolved organic carbon (DOC), prokaryotic cell count and 32 chemicals (ammonium, nitrites, nitrates, heavy metals, inorganic pollutants, PAHs (polycyclic aromatic hydrocarbons), pesticides and organochlorines). None of the tested chemicals were detected at concentrations higher than the European and national threshold values. Dissolved organic carbon (DOC) was 1.1 mg/L. The collection of copepods was performed in June 2017 when the temperature of groundwater was 15.3 °C, i.e., very close to the mean annual value (15 °C).

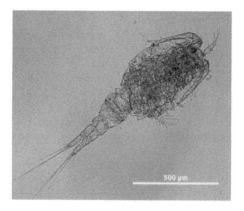

Figure 1. Female of *Diacyclops belgicus* at the optical microscope.

A phreatobiological net-sampler (mesh size: 60 μm) was used to collect copepods from the bottom and the water column of the well. After the collection, the samples were transferred to the laboratory within 10 min. In the laboratory, the individuals of *D. belgicus* were picked up by a glass pipette and pooled in a 100-mL glass vial filled with the groundwater collected from the well (hereafter called bore water). The culture was kept in the dark in a thermostatic cabinet (Mod. ST 3, Pol-Eko-Aparatura, Wodzisław Śląski, Poland) at 14 °C for 7 days. The bore water was not renewed and no additional food was offered. Microbes occurred with about 1.8×10^6 prokaryotic cells/mL in the bore water (Flow Cytometer A50-micro, Apogee Flow System, Hertfordshire, UK). Eighteen hours before the trial at 14 °C the copepods were rinsed in standard water (1 L of MILLIPORE MILLI-Q® (Elix®, Merck KGaA, Darmstadt, Germany) deionized water remineralized with the following reagent grade chemicals: 0.06 g of $MgSO_4$, 0.096 g of $NaHCO_3$, 0.004 g of KCl, 0.06 g of $CaSO_4 \cdot 2H_2O$; [25]) and kept in a 10-mL glass vial with standard water and in the darkness at the testing temperature. No food was offered during acclimation in order to allow gut-emptying [26,27]. After the acclimation, only the actively swimming individuals were selected for testing. The test individuals (juvenile stages, i.e., copepodids) were picked up by a glass pipette under a Leica M80 stereomicroscope at 20× magnification. The pick-up of fecal pellets was carefully avoided so to exclude the overshoot in oxygen consumption due to digestive metabolism [20,26,27]. The respirometric measurements were started on the day 8th for the trials at 14 °C. The whole procedure (collection of new animals from the field, acclimation in bore water and in standard water and testing) was repeated for the trial at 17 °C at the day 9th. A sampling effect on the trials was not considered likely since groundwater habitats have a strong physical inertia on a monthly scale [11–15]. The animals that had been not used in the tests were kept in the bore water in the thermostatic cabinet.

2.2. Measurement of Oxygen Consumption

Standard respiration rates (SRRs) were measured in a sealed glass microplate equipped with planar oxygen sensor spots with optical isolation glued onto the bottom of 80-µL wells (Loligo Systems, Viborg, Denmark). The microplate was integrated with a 24-channel fluorescence-based respirometry system (SDR Sensor Dish Reader) (PreSens Precision Sensing GmbH, Regensburg, Germany) that allows simultaneous measurement of 20 replicates and 4 controls (wells without animals). This respirometer is known for its simplicity, high throughput, and high temporal resolution and sensitivity [28,29]. The reader was placed inside the thermostatic cabinet at the appropriate testing temperatures (either 14 or 17 °C; accuracy: ±0.1 °C) 18 h before the beginning of the trials to bring the equipment and the standard water to the same temperature prior to measurements [20,29]. Twenty copepodids of *D. belgicus* were then individually loaded in the microwells that were in turn inspected for air bubbles under a stereomicroscope (Leica M80, Leica Microsystems Srl, Wetzlar, Germany) at 20× magnification. The microwells were overfilled with oversaturated standard water and sealed with parafilm before the plate was placed on the SDR reader. The well inspection was repeated after the sealing with the parafilm. Four microwells were kept without animals for controls and they were filled with the standard water used for copepod acclimation. A 10 cm × 7 cm × 0.5 cm rectangular silicone layer was put on top of the microplate for a further sealing. Finally, a weight was posed on top of everything. The preparation of the microplate was conducted under the stereomicroscope at room temperature within 20 min. Nevertheless, when the sealed microplate was put back into the thermostatic cabinet after copepod loading, the device temperature was always about 2 °C higher than the respective test temperature (15.9 °C vs. 14 °C; 18.8 °C vs. 17 °C). The measurements were started as soon as the device temperature had reached the equilibrium with the test temperature, by the SDR v4 Software (PreSens Precision Sensing GmbH, Regensburg, Germany). From preliminary experiments, we set at 120 min the time needed to temperature re-equilibration. Afterwards, the oxygen consumption (% air saturation) was recorded every 2 min during 2 h after the temperature re-equilibration. The plate was not shaken allowing the animals to respire without stress. Total oxygen consumption (µg O_2 per individual per hour) was calculated while correcting for observed changes in oxygen readings in the control wells (mean values of the 4 controls). The percentage of air saturation was above 80% in each well at the end of the trials at both test temperatures.

At the end of each trial, the microwells were opened and inspected under the microscope at 50× magnification. The copepods were individually picked-up by a glass pipette and loaded in a Petri dish and checked for mortality (no movement after gentle stimulation by means of a sorting needle). No dead individuals were detected at the end of the trials. Some drops of carbonated water were added to the Petri dish in order to narcotize the individuals. Pictures of each individual were taken by a HD camera (MC 170, Leica Microsystems Srl, Wetzlar, Germany) that was integrated to the microscope. The camera was connected to a notebook and the pictures were elaborated by LAS EZ vs. 3 software (Leica Microsystems Srl, Wetzlar, Germany) and the free vector graphic software Inkscape (https://inkscape.org/it/). The prosome length (L, mm) and width (W, mm) of each individual were measured using the methods described in our previous studies [26,27]. Dry mass was calculated according to [30]:

$$\text{wet mass (WM)} = K \times (\text{prosomal length}) \times \text{width}^2 \tag{1}$$

where K is a constant, being 0.705 for cyclopoids [31]. A conversion factor of 0.25 was used to convert the wet mass to the dry mass [32], i.e.,:

$$\text{dry mass (DM)} = 0.25 \times \text{WM.} \tag{2}$$

To describe the relationship between temperature and performance we used the metric Q_{10} that is the rate of change of a biological system (in this case, oxygen consumption) as a consequence of a temperature increase of 10 °C. This value was calculated between 14 and 17 °C using the formula:

$$Q_{10} = (R_2/R_1)^{10/(T_2-T_1)} \tag{3}$$

where R_1 and R_2 are the oxygen consumptions at temperatures T_1 and T_2 and $T_2 > T_1$ [20,33]. We used both the mean and the median values of the oxygen consumption.

2.3. Statistical Analyses

To test for differences in size (L, W and DM) between the two trials (at 14 and 17 °C) we used three different one-way permutational analyses of variance (PERMANOVA) [34], one per each variable, with the factor "temperature" (two levels: 14 and 17 °C). To test for differences in the oxygen consumption in µg O_2 per individual per hour, we used a one-way permutational analysis of covariance (PERMANCOVA) [34], with individual DM as the covariate. The permutational analyses were performed on the basis of a Euclidean Distance similarity matrix and using either non-transformed or log(x + 1) transformed data, after performing a Levene's test on the original dataset to check for variance homogeneity. PERMANOVA was run under Type III Sum of Squares and unrestricted permutation of raw data. PERMANCOVA was run under Type I Sum of Squares and permutation of residuals under a reduced model [34]. The number of permutations required to derive p-values was set at 9999. Whenever the number of unique permutations was lower than 100, Monte Carlo p-values were preferred over the permutation p-values.

Prior to the analyses, all data were checked for outliers based on the following fences: lower inner fence (Q1 − 1.5 × IQ), upper inner fence (Q3 + 1.5 × IQ), lower outer fence (Q1 − 3 × IQ) and upper outer fence (Q3 + 3 × IQ), where Q1 and Q3 are the first and the third quartiles respectively and IQ is the interquartile range (i.e., Q3 − Q1). We assumed as outliers those data that were either lower than the lower fences or higher than the upper fences. PERMANOVAs and PERMANCOVA were run on both the raw data and on the raw data set deprived of outliers.

All the statistical tests were performed using PRIMER v.6 & PERMANOVA + routines for PRIMER (PRIME-E, Auckland, New Zealand) [35]. The level of significance was set to $p < 0.05$.

3. Results

SRR, L, W and DM of the individuals used in the trials at 14 and 17 °C are shown in Table 1. L, W and DM of the individuals used in the trial at 14 °C were not significantly different from those at 17 °C (PERMANOVA; L: Pseudo-$F_{1,38}$ = 1.06, p-value = 0.3074, perms = 9834; W: Pseudo-$F_{1,38}$ = 0.45, MonteCarlo p-value = 0.5269, perms = 67; DM: Pseudo-$F_{1,38}$ = 0.003, p-value = 0.9580, perms = 9823). The results of the PERMANCOVA showed that there was not a significant relationship between the dry mass and the SRR of the individuals (PERMANCOVA; DM: Pseudo-$F_{1,36}$ = 1.74, p-value = 0.1941, perms = 9817). Nevertheless, even given the non-occurrence of variation of the SRRs due to the dry mass, no significant variability was detected between the trials at 14 and 17 °C (PERMANCOVA; SRR: Pseudo-$F_{1,36}$ = 1.82, p-value = 0.1878, perms = 9839).

Table 1. Prosome length (L), prosome width (W), dry mass (DM) and standard respiration rates (SRR) of the 40 individuals of *Diacyclops belgicus* (Kiefer, 1936) involved in the respirometric trials at two temperatures (T). Mean, standard deviation (SD) and minimum and maximum values are also shown. SRR are in µg of oxygen per individual per hour. Outliers above the upper inner fence are in bold and underlined.

Microwell	T (°C)	L (mm)	W (mm)	DM (mg)	SRR (µg O$_2$/ind. ×h)
W1	14	0.2521	0.1123	0.0006	0.0030
W2	14	0.2356	0.1151	0.0006	0.0030
W3	14	0.2630	0.0959	0.0004	0.0148
W4	14	0.2356	0.0877	0.0003	0.0131
W5	14	0.2795	0.0904	0.0004	0.0059
W6	14	0.3069	0.1206	0.0008	0.0089
W7	14	0.2274	0.1014	0.0004	0.0148
W8	14	0.2795	0.0959	0.0005	0.0059
W9	14	0.3206	0.1452	**_0.0012_**	0.0119
W10	14	0.2411	0.1123	0.0005	0.0163
W11	14	0.2795	0.0904	0.0004	0.0074
W12	14	0.2713	0.0795	0.0003	0.0074
W13	14	0.2384	0.0932	0.0004	0.0074
W14	14	0.2987	0.0849	0.0004	0.0074
W15	14	0.2356	0.1151	0.0006	0.0030
W16	14	0.2795	0.0904	0.0004	0.0059
W17	14	0.3206	0.1452	**_0.0012_**	0.0119
W18	14	0.2411	0.1123	0.0005	0.0163
W19	14	0.2713	0.0795	0.0003	0.0074
W20	14	0.2987	0.0849	0.0004	0.0074
W1	17	0.2411	0.1233	0.0006	0.0037
W2	17	0.2521	0.1206	0.0006	0.0120
W3	17	0.2630	0.1178	0.0006	0.0037
W4	17	0.2411	0.1096	0.0005	0.0037
W5	17	0.2630	0.1151	0.0006	**_0.0232_**
W6	17	0.2658	0.1233	0.0007	0.0037
W7	17	0.2302	0.1069	0.0005	0.0037
W8	17	0.2795	0.1233	0.0007	0.0042
W9	17	0.2822	0.0986	0.0005	0.0014
W10	17	0.2576	0.0986	0.0004	0.0014
W11	17	0.2576	0.1014	0.0005	0.0014
W12	17	0.2685	0.1014	0.0005	0.0065
W13	17	0.2822	0.0904	0.0004	0.0037
W14	17	0.2548	0.1069	0.0005	0.0148
W15	17	0.2493	0.0932	0.0004	0.0148
W16	17	0.3014	0.1233	0.0008	0.0148
W17	17	0.2767	0.1041	0.0005	0.0037
W18	17	0.2740	0.0986	0.0005	0.0093
W19	17	0.2439	0.0932	0.0004	0.0037
W20	17	0.2192	0.0740	0.0002	0.0015

Statistics	T (°C)	L (mm)	W (mm)	DM (mg)	SRR (µg O$_2$/ind. ×h)
Mean	14	0.2688	0.1026	0.0005	0.0090
Mean	17	0.2602	0.1062	0.0005	0.0067
SD	14	0.0298	0.0193	0.0003	0.0044
SD	17	0.0198	0.0135	0.0001	0.0060
Max	14	0.3206	0.1452	0.0012	0.0163
Max	17	0.3014	0.1233	0.0008	0.0232
Min	14	0.2274	0.0795	0.0003	0.0030
Min	17	0.2192	0.0740	0.0002	0.0014

Two dry mass and 1 SRR values were above the respective upper inner fences and were thus excluded from the analyses (Table 1). The dry masses of the individuals used in the trial at

14 °C were not significantly different from those at 17 °C (PERMANOVA; DM: Pseudo-$F_{1,36}$ = 3.20, p-value = 0.0805, perms = 9842). The results of the PERMANCOVA showed that there was not a significant relationship between the dry mass and the SRRs of the individuals (PERMANCOVA; DM: Pseudo-$F_{1,33}$ = 9.91, p-value = 0.7584, perms = 9835). In addition, no significant variability in the SRRs was detected between the trials at 14 and 17 °C (PERMANCOVA; SRR: Pseudo-$F_{1,33}$ = 3.77, p-value = 0.0616, perms = 9853).

No outliers higher or lower the outer fences were found.

Q_{10} was 0.39 considering the mean values of the oxygen consumptions and 0.99 considering the median values.

4. Discussion

Groundwater ecosystems are generally poorer in nutrients and oxygen than surface water ecosystems [12–15,36–40]. In order to reduce energetic costs, groundwater ectotherms have evolved metabolic rates that are lower than those of their close epigean relatives [20,41–44]. The results of the one and only study on the metabolism of a stygobitic copepod species [26], in which we compared the SRRs of the stygobitic *D. belgicus* and the epigean *Eucyclops serrulatus* (belonging to the same family Cyclopidae), were consistent with this statement. The SRRs of *E. serrulatus* was 7 and 5 fold the SRRs of *D. belgicus* juveniles and adults, respectively. Albeit measured with a different device and protocol and under different temperatures, the individual-based measurements of the SRRs of *D. belgicus* of the present study are of the same order of magnitude (nanograms of O_2 per individual per hour) as those observed in [26]. Details are provided in Table S1. The scarcity of available data and the difficulty in performing experiments with stygobitic species have frequently led to the sensitivity of groundwater taxa to environmental stressors being inferred from data obtained for epigean relatives. This approach has resulted in the assessment of threshold values for groundwater ecosystem quality that are not protective of the biota which live in this environment [14,45,46]. There are severe constraints that make difficult to perform metabolic, physiological and ecotoxicological studies with groundwater microcrustaceans in laboratory, most of them related to the low metabolism of these species. To our knowledge, no researcher has determined how to make groundwater copepods reproduce in the laboratory. We ourselves kept cultures of *D. belgicus* in laboratory for one year before this study and we did not observe any ovigerous female in the cohorts. This limits the number of replicates and sometimes prevents performing a trial. Most importantly, the abundances are very low and "the one well to one species" is the rule of thumb [12,14,15,40].

Very few studies have investigated the importance of temperature variations for obligate groundwater species [19,20,23]. In this study, the copepod *D. belgicus* did not significantly varied its SRRs between 14 and 17 °C after 7-days acclimation at the testing temperatures. During exposure to a range of temperatures, the relationship between the body temperature and metabolism can be described by an asymmetric bell-shaped function, where the metabolic rate is maximized at an intermediate temperature: the thermal optimum [18,47]. As found for many other species [20,48–52], acclimation to temperatures that are different from the thermal optimum has a significant impact on crustaceans physiology that may change in order to confer protection against the injuries produced by temperature variation. Under acclimation, hemolymph pH may vary [53], as well as enzyme properties [54] and hemoglobin affinities [55], as response to temperature variations. In this study, we did not measure specifically these physiological changes, however we used the respirometric metabolism as a proxy. Nevertheless, the SRRs of *D. belgicus* at 17 °C were not significantly different from those at 14 °C thus indicating a non-significant variation in the physiological processes. Similarly, the SRRs at 14, 17 and 21 °C of the stygobitic amphipods *N. renorhodanensis* and *N. virei* were very close to each other, although differences had been not tested by statistical analyses [20]. Moreover, the mode of the bell-shaped curve of oxygen consumption and temperature fell around 17 and 21 °C for *N. rhenorhodanensis* which consequently has led it to be characterized as a eurythermal species [20]. Accordingly, individuals of a population of the stygobitic isopod *Proasellus valdensis* (Chappuis, 1948) which survived in the laboratory at temperatures up to 22 °C, no longer increased their metabolic rates above 16 °C [23].

The thermal tolerance of *D. belgicus* is relevant to the habitat where this species was collected for this study, that is a Mediterranean shallow aquifer characterized by annual temperature amplitude of about 2 °C. The individuals of *D. belgicus* that reside in the Medio Valdarno porous aquifer are adapted to temperatures that vary between 14 and 15.9 °C and do not change its respiratory metabolism under a further increasing of 1.1 °C, at least as far as the juvenile stages are concerned and under 7-days acclimation. As high metabolic rates are expected to increase the influx rate of toxicants in aquatic crustaceans [27,44,56], the invariability of the SRRs between 14 and 17 °C may be considered protective to this species. In addition, the absence of SRR variability suggests that the survival of *D. belgicus* at a temperature of 1.1 °C above the highest measured in the Medio Valdarno porous aquifer does not depend on the ability of this species to satisfy its aerobic metabolic demand [57]. This is clearly an advantage in oligotrophic environments.

Acclimation timing also matters. An organism might easily cope with an individual short-term heat shock but be compromised if the same shock is applied repeatedly or continuously. On the other hand, repeated sub-lethal heat shocks can lead to a form of acclimation in which the body's tolerance becomes more robust [58]. Thus, we cannot exclude the possibility that an early increase in the respirometric metabolism of *D. belgicus* might occur few hours after the acclimation and be recovered after some days.

The Q_{10} of *D. belgicus* between 14 and 17 °C was 0.39 considering the mean value of oxygen consumption. Q_{10} values between 21 and 28 °C were also smaller than 1 for the amphipod *Gammarus fossarum* Koch, 1836 ($Q_{10} = 0.66$), *N. rhenorhodanensis* ($Q_{10} = 0.68$) and *N. virei* ($Q_{10} = 0.36$) [20]. $Q_{10} < 1$ indicates that a temperature change of 10 °C respect to the thermal optimum is likely to damage the metabolic system of these species leading to what may be an irreversible loss of functions [33]. It also suggests a low capacity to maintain optimal enzymatic activities and a limited survival as soon as the environmental temperature goes out of the optimal thermal range of the species. An irreversible denaturation of enzymes/proteins seems to begin for *D. belgicus* between 14 and 17 °C according to the $Q_{10} = 0.39$. However, the Q_{10} metric is computed with the mean of the SRR values whose partition between 14 and 17 °C did not differ significantly. In addition, when computed with the median SRR values, the Q_{10} is 0.99 for *D. belgicus* and a Q_{10} of approximately 1, supports optimal physiological performance and thermal tolerance. The wide geographical range of this species appears to support the thermal tolerance hypothesis [59], which likely favors dispersal, especially via the porous medium. However, further experiments at temperatures above and below those tested in this study are required to assess the optimal thermal range of this species.

5. Conclusions

The results of this study derive from an experimental increase of temperature that is consistent with the foreseen increase in the next 30 years due to global warming. The populations of *D. belgicus* of the Medio Valdarno porous aquifer did not show significant variations in the oxygen consumptions under a temperature change of 3 °C. This result is in agreement with the wide geographical range of occurrence of this species. The result is supported by a thermal coefficient Q_{10} ~1, computed on the median values of the oxygen consumptions, that suggests a thermal tolerance of this species. Conversely, the value of the thermal coefficient Q_{10} was far below 1 when the mean values of the oxygen consumptions were considered. This value indicates the probable beginning of an irreversible denaturation of enzymes/proteins at a temperature increase of 10 °C from the thermal optimum. These controversial results do not provide full certainty about the fate of this species under a global warming scenario. Hence, *D. belgicus* should be prudently considered at risk. Further experiments at temperatures above and below those tested in this study are required to assess the optimal thermal range of *D. belgicus*. In addition, new studies are required with groundwater species exhibiting narrow distributions. The physiological capacity to respond to temperature changes is expected to be very low in narrowly distributed species, thus facing a higher risk of extinction due to global warming.

Supplementary Materials: The following are available online at www.mdpi.com/2073-4441/9/12/951/s1.

Acknowledgments: The project, including the costs to publish in open access, was funded by the European Community (LIFE12 BIO/IT/000231 AQUALIFE). We thank Giuseppe Messana (ISE-CNR) and Sigurd Einum (NTNU) for the tips for the use of the respirometric device, Claudio Sili (ISE-CNR) for the picture of *D. belgicus* and Stefano Amalfitano (IRSA-CNR) for the flow cytometry. We would like to thank three anonymous reviewers who provided comments to improve the manuscript.

Author Contributions: Both authors conceived and designed the experiments; Tiziana Di Lorenzo performed the experiments; both authors analyzed the data and wrote the paper.

Conflicts of Interest: The authors declare no conflict of interest. The founding sponsors had no role in the design of the study; in the collection, analyses, or interpretation of data; in the writing of the manuscript, and in the decision to publish the results.

References

1. Gamvroudis, C.; Dokou, Z.; Nikolaidis, N.P.; Karatzas, G.P. Impacts of surface and groundwater variability response to future climate change scenarios in a large Mediterranean watershed. *Environ. Earth Sci.* **2017**, *76*. [CrossRef]

2. Dettinger, M.D.; Earman, S. Western Ground Water and Climate Change—Pivotal to Supply Sustainability or Vulnerable in Its Own Right? *Ground Water* **2007**, *4*. Available online: http://tenaya.ucsd.edu/~dettinge/agwse07.pdf (accessed on 23 October 2017).

3. Taylor, R.G.; Scanlon, B.; Döll, P.; Rodell, M.; van Beek, R.; Wada, Y.; Longuevergne, L.; Leblanc, M.; Famiglietti, J.S.; Edmunds, M.; et al. Ground water and climate change. *Nat. Clim. Chang.* **2007**, *3*, 322–329. [CrossRef]

4. Intergovernmental Panel on Climate Change (IPCC). The physical science basis. In *Contribution of Working Group I to the Fourth Assessment Report of the Intergovernmental Panel on Climate Change*; Solomon, S., Qin, D., Manning, M., Chen, Z., Marquis, M., Averyt, K.B., Tignor, M., Miller, H.L., Eds.; Cambridge University Press: Cambridge, UK; New York, NY, USA, 2007.

5. Kundzewicz, Z.W.; Mata, L.J.; Arnell, N.W.; Doll, P.; Kabat, P.; Jimenez, B.; Miller, K.A.; Oki, T.; Sen, Z.; Shiklomanov, I.A. Freshwater resources and their management. In *Climate Change 2007: Impacts, Adaptation and Vulnerability*; Parry, M.L., Canziani, O.F., Palutikof, J.P., van der Linden, P.J., Hanson, C.E., Eds.; Cambridge University Press: Cambridge, UK; New York, NY, USA, 2007; pp. 173–210.

6. Green, T.R.; Taniguchi, M.; Kooi, H.; Gurdakd, J.J.; Allen, D.M.; Hiscock, K.M.; Treidel, H.; Aurelig, A. Beneath the surface of global change: Impacts of climate change on groundwater. *J. Hydrol.* **2011**, *405*, 532–560. [CrossRef]

7. Intergovernmental Panel on Climate Change (IPCC). Climate Change (IPCC). Climate Change 2013: The physical science basis. In *Contribution of Working Group I to the fifth Assessment Report of the Intergovernmental Panel On Climate Change*; Stocker, T.F., Qin, D., Plattner, G.-K., Tignor, M.M.B., Allen, S.K., Boschung, J., Nauels, A., Xia, Y., Bex, V., Midgley, P.M., Eds.; Cambridge University Press: Cambridge, UK; New York, NY, USA, 2014. [CrossRef]

8. Menberg, K.; Blum, P.; Kurylyk, B.L.; Bayer, P. Observed groundwater temperature response to recent climate change. *Hydrol. Earth Syst. Sci.* **2014**, *18*, 4453–4466. [CrossRef]

9. Figura, S.; Livingstone, D.M.; Hoehn, E.; Kipfer, R. Regime shift in groundwater temperature triggered by the Arctic Oscillation. *Geophys. Res. Lett.* **2011**, *38*, L23401. [CrossRef]

10. Taylor, A.C.; Stefan, H.G. Shallow groundwater temperature response to climate change and urbanization. *J. Hydrol.* **2009**, *375*, 601–612. [CrossRef]

11. Freeze, R.A.; Cherry, J.A. *Groundwater*; Prentice-Hall, Inc.: Englewood Cliffs, NJ, USA, 1979; p. 604.

12. Galassi, D.M.P.; Stoch, F.; Fiasca, B.; Di Lorenzo, T.; Gattone, E. Groundwater biodiversity patterns in the Lessinian Massif of northern Italy. *Freshwat. Biol.* **2009**, *54*, 830–847. [CrossRef]

13. Di Lorenzo, T.; Brilli, M.; Del Tosto, D.; Galassi, D.M.P.; Petitta, M. Nitrate source and fate at the catchment scale of the Vibrata River and aquifer (central Italy): An analysis by integrating component approaches and nitrogen isotopes. *Environ. Earth Sci.* **2012**. [CrossRef]

14. Di Lorenzo, T.; Cifoni, M.; Lombardo, P.; Fiasca, B.; Galassi, D.M.P. Ammonium threshold value for groundwater quality in the EU may not protect groundwater fauna: Evidence from an alluvial aquifer in Italy. *Hydrobiologia* **2015**, *743*, 139–150. [CrossRef]

15. Di Lorenzo, T.; Galassi, D.M.P. Agricultural impact on Mediterranean alluvial aquifers: Do groundwater communities respond? *Fundam. Appl. Limnol.* **2013**, *182*, 271–282. [CrossRef]

16. Eckert, R.; Randall, D.; Burggren, W.; French, K. *Animal Physiology: Mechanisms and Adaptations*; Freeman and Company: New York, NY, USA, 1979; p. 120, ISBN 10:0716738635.

17. Peck, L.S.; Webb, K.E.; Bailey, D.M. Extreme sensitivity of biological function to temperature in Antarctic marine species. *Funct. Ecol.* **2004**, *18*, 625–630. [CrossRef]

18. Huey, R.B.; Kingsolver, J.G. Evolution of thermal sensitivity of ectotherm performance. *Trends Ecol. Evol.* **1989**, *4*, 131–135. [CrossRef]

19. Issartel, J.; Renault, D.; Voituron, Y.; Bouchereau, A.; Vernon, P.; Hervant, F. Metabolic responses to cold in subterranean crustaceans. *J. Exp. Biol.* **2005**, *208*, 2923–2929. [CrossRef] [PubMed]

20. Issartel, J.; Hervant, F.; Voituron, Y.; Renault, D.; Vernon, P. Behavioural, ventilatory and respiratory responses of epigean and hypogean crustaceans to different temperatures. *Comp. Biochem. Physiol.* **2005**, *141*, 1–7. [CrossRef] [PubMed]

21. Colson-Proch, C.; Renault, D.; Gravot, A.; Douady, C.J.; Hervant, F. Do current environmental conditions explain physiological and metabolic responses of subterranean crustaceans to cold? *J. Exp. Biol.* **2009**, *212*, 1859–1868. [CrossRef] [PubMed]

22. Colson-Proch, C.; Morales, A.; Hervant, F.; Konecny, L.; Moulin, C.; Douady, C.J. First cellular approach of the effects of global warming on groundwater organisms: A study of the HSP70 gene expression. *Cell Stress Chaperon.* **2010**, *15*, 259–270. [CrossRef] [PubMed]

23. Mermillod-Blondin, F.; Lefour, C.; Lalouette, L.; Renault, D.; Malard, F.; Simon, L.; Douady, C.J. Thermal tolerance breadths among groundwater crustaceans living in a thermally constant environment. *J. Exp. Biol.* **2013**, *216*, 1683–1694. [CrossRef] [PubMed]

24. Pesce, L. The genus *Diacyclops* Kiefer in Italy: A taxonomic, ecological and biogeographical up-to-date review (Crustacea Copepoda Cyclopidae). *Arthropoda Sel.* **1994**, *3*, 13–19.

25. Cifoni, M.; Galassi, D.M.P.; Faraloni, C.; Di Lorenzo, T. Test procedures for measuring the (sub)chronic effects of chemicals on the freshwater cyclopoid *Eucyclops serrulatus*. *Chemosphere* **2017**, *173*, 89–98. [CrossRef] [PubMed]

26. Di Lorenzo, T.; Di Marzio, W.D.; Spigoli, D.; Baratti, M.; Messana, G.; Cannicci, S.; Galassi, D.M.P. Metabolic rates of a hypogean and an epigean species of copepod in an alluvial aquifer. *Freshw. Biol.* **2015**, *60*, 426–435. [CrossRef]

27. Di Lorenzo, T.; Cannicci, S.; Spigoli, D.; Cifoni, M.; Baratti, M.; Galassi, D.M.P. Bioenergetic cost of living in polluted freshwater bodies: Respiration rates of the cyclopoid *Eucyclops serrulatus* under ammonia-N exposures. *Fundam. Appl. Limnol.* **2016**, *188*, 147–156. [CrossRef]

28. Szela, T.L.; Marsh, A.G. Microtiter plate, optode respirometry, and inter-individual variance in metabolic rates among nauplii of *Artemia* sp. *Mar. Ecol. Prog. Ser.* **2005**, *296*, 281–289. [CrossRef]

29. Yashchenko, V.; Fossen, E.I.; Kielland, Ø.N.; Einum, S. Negative relationships between population density and metabolic rates are not general. *J. Anim. Ecol.* **2016**, *85*, 1070–1077. [CrossRef] [PubMed]

30. Svetlichny, L.S.; Khanaychenko, A.; Hubareva, E.; Aganesova, L. Partitioning of respiratory energy and environmental tolerance in the copepods *Calanipeda aquaedulcis* and *Arctodiaptomus salinus*. *Estuar. Coast. Shelf. Sci.* **2012**, *114*, 199–207. [CrossRef]

31. McKinnon, A.D.; Duggan, S. Summer copepod production in subtropical waters adjacent to Australia's North West Cape. *Mar. Biol.* **2003**, *143*, 897–907. [CrossRef]

32. Reiss, J.; Schmid-Araya, J.M. Feeding response of a benthic copepod to ciliate prey type, prey concentration and habitat complexity. *Freshw. Biol.* **2011**, *56*, 1519–1530. [CrossRef]

33. Hochachka, P.; Somero, G. *Biochemical Adaptation, Mechanism and Physiological Evolution*; Oxford University Press: New York, NY, USA, 2002; p. 480, ISBN 9780195117035.

34. Anderson, M.J. A new method for non-parametric multivariate analysis of variance. *Austral Ecol.* **2001**, *26*, 32–46. [CrossRef]

35. Anderson, M.J.; Gorley, R.N.; Clarke, K.R. *PERMANOVA + for PRIMER: Guide to Software and Statistical Methods*; PRIMER-E Ltd.: Plymouth, UK, 2008.

36. Galassi, D.M.P.; Lombardo, P.; Fiasca, B.; Di Cioccio, A.; Di Lorenzo, T.; Petitta, M.; Di Carlo, P. Earthquakes trigger the loss of groundwater biodiversity. *Sci. Rep.* **2014**, *4*. [CrossRef] [PubMed]

37. Galassi, D.M.P.; Fiasca, B.; Di Lorenzo, T.; Montanari, A.; Porfirio, S.; Fattorini, S. Groundwater biodiversity in a chemoautotrophic cave ecosystem: How geochemistry regulates microcrustacean community structure. *Aquat. Ecol.* **2017**, *51*, 75–90. [CrossRef]

38. Di Lorenzo, T.; Stoch, F.; Galassi, D.M.P. Incorporating the hyporheic zone within the river discontinuum: Longitudinal patterns of subsurface copepod assemblages in an Alpine stream. *Limnologica* **2013**, *43*, 288–296. [CrossRef]

39. Stoch, F.; Barbara, F.; Di Lorenzo, T.; Porfirio, S.; Petitta, M.; Galassi, D.M.P. Exploring copepod distribution patterns at three nested spatial scales in a spring system: Habitat partitioning and potential for hydrological bioindication. *J. Limnol.* **2016**, *75*, 1–13. [CrossRef]

40. Iepure, S.; Rasines-Ladero, R.; Meffe, R.; Carreno, F.; Mostaza, D.; Sundberg, A.; Di Lorenzo, T.; Barroso, J.L. The role of groundwater crustaceans in disentangling aquifer type features—A case study of the Upper Tagus Basin, central Spain. *Ecohydrology* **2017**, *10*, e1876. [CrossRef]

41. Hervant, F.; Mathieu, J.; Messana, G. Locomotory, ventilatory and metabolic responses of the subterranean *Stenasellus virei* (Crustacea, Isopoda) to severe hypoxia and subsequent recovery. *C. R. Acad. Sci.* **1997**, *320*, 139–148. [CrossRef]

42. Hervant, F.; Renault, D. Long-term fasting and realimentation in hypogean and epigean isopods: A proposed adaptive strategy for groundwater organisms. *J. Exp. Biol.* **2002**, *205*, 2079–2087. [PubMed]

43. Simčič, T.; Lukančič, S.; Brancelj, A. Comparative study of electron transport system activity and oxygen consumption of amphipods from caves and surface habitats. *Freshw. Biol.* **2005**, *50*, 494–501. [CrossRef]

44. Simčič, T.; Pajk, F.; Brancelj, A. Electron transport system activity and oxygen consumption of two amphibious isopods, epigean *Ligia italic* Fabricius and hypogean *Titanethes albus* (Koch), in air and water. *Mar. Freshw. Behav. Physiol.* **2010**, *43*, 149–156. [CrossRef]

45. Di Lorenzo, T.; Di Marzio, W.D.; Sáenz, M.E.; Baratti, M.; Dedonno, A.A.; Iannucci, A.; Cannicci, S.; Messana, G.; Galassi, D.M.P. Sensitivity of hypogean and epigean freshwater copepods to agricultural pollutants. *Environ. Sci. Pollut. Res.* **2014**, *21*, 4643–4655. [CrossRef] [PubMed]

46. Di Lorenzo, T.; Di Marzio, W.D.; Cifoni, M.; Fiasca, B.; Baratti, M.; Sáenz, M.E.; Galassi, D.M.P. Temperature effect on the sensitivity of the copepod *Eucyclops serrulatus* (Crustacea, Copepoda, Cyclopoida) to agricultural pollutants in the hyporheic zone. *Curr. Zool.* **2015**, *61*, 629–640. [CrossRef]

47. Angilletta, M.J.; Niewiarowski, P.H.; Navas, C.A. The evolution of thermal physiology in ectotherms. *J. Therm. Biol.* **2002**, *27*, 249–268. [CrossRef]

48. Holmstrup, M.; Costanzo, J.P.; Lee, R.E., Jr. Cryoprotective and osmotic responses to cold acclimation and freezing in freeze-tolerant and freeze-intolerant earthworms. *J. Comp. Physiol.* **1999**, *169*, 207–214. [CrossRef]

49. Bale, J.S.; Block, W.; Worland, M.R. Thermal tolerance and acclimation response of larvae of the sub-Antarctic beetle *Hydromedion sparsutum* (Coleoptera: Perimylopidae). *Polar Biol.* **2000**, *23*, 77–84. [CrossRef]

50. Renault, D.; Vernon, P.; Nedved, O.; Hervant, F. The importance of fluctuating thermal regimes for repairing chill injuries in the tropical beetle *Alphitobius diaperinus* (Coleoptera: Tenebrionidae) during exposure to low temperature. *Physiol. Entomol.* **2004**, *29*, 139–145. [CrossRef]

51. Simčič, T.; Brancelj, A. Electron transport system (ETS) activity in five *Daphnia* species at different temperatures. *Hydrobiologia* **1997**, *360*, 117–125. [CrossRef]

52. Simčič, T.; Pajk, F.; Vrezec, A.; Brancelj, A. Size scaling of whole-body metabolic activity in the noble crayfish (*Astacus astacus*) estimated from measurements on a single leg. *Freshw. Biol.* **2012**, *57*, 39–48. [CrossRef]

53. Tanaka, K.; Udagawa, T. Cold adaptation of the terrestrial isopod, *Porcellio scaber*, to subnivean environments. *J. Comp. Physiol.* **1993**, *163*, 439–444.

54. Mulkiewicz, E.; Zietara, M.S.; Stachowiak, K.; Skorkowski, E.F. Properties of lactate dehydrogenase from the isopod, *Saduria entomon*. *Comp. Biochem. Physiol.* **2000**, *126*, 337–346. [CrossRef]

55. Paul, R.J.; Lamkemeyer, T.; Maurer, J.; Pinkhaus, O.; Pirow, R.; Seidl, M.; Zeis, B. Thermal acclimation in the microcrustacean *Daphnia*: A survey of behavioural, physiological and biochemical mechanisms. *J. Therm. Biol.* **2004**, *29*, 655–662. [CrossRef]

56. Gutierrez, F.M.; Gagneten, A.M.; Paggi, J.C. Copper and chromium alter life cycle variables and the equiproportional development of the freshwater copepod *Notodiaptomus conifer* (SARS). *Water Air Soil Pollut.* **2010**, *213*, 275–286. [CrossRef]

57. Pörtner, H.-O. Climate variations and the physiological basis of temperature dependent biogeography: Systemic to molecular hierarchy of thermal tolerance in animals. *Comp. Biochem. Physiol.* **2002**, *132*, 739–761. [CrossRef]

58. Dowd, W.W.; King, F.A.; Denny, M.W. Thermal variation, thermal extremes and the physiological performance of individuals. *J. Exp. Biol.* **2015**, *218*, 1956–1967. [CrossRef] [PubMed]

59. Eme, D.; Malard, F.; Colson-Proch, C.; Jean, P.; Calvignac, S.; Konecny-Dupr, L.; Hervant, F.; Douady, C.J. Integrating phylogeography, physiology and habitat modelling to explore species range determinants. *J. Biogeogr.* **2014**, *41*, 687–699. [CrossRef]

Article

Distribution Patterns of the Freshwater Oligochaete *Limnodrilus hoffmeisteri* Influenced by Environmental Factors in Streams on a Korean Nationwide Scale

Hyejin Kang [1], Mi-Jung Bae [2], Dae-Seong Lee [1], Soon-Jin Hwang [3], Jeong-Suk Moon [4] and Young-Seuk Park [1,5,*]

1   Department of Biology, Kyung Hee University, Dongdaemun-gu, Seoul 02447, Korea;
     ghj2166@naver.com (H.K.); dleotjd520@naver.com (D.-S.L.)
2   Freshwater Biodiversity Research Division, Nakdonggang National Institute of Biological Resources,
     Sangju, Gyeongsangbuk-do 37242, Korea; mjbae@nnibr.re.kr
3   Department of Environmental Health Science, Konkuk University, Gwangjin-gu, Seoul 05029, Korea;
     sjhwang@konkuk.ac.kr
4   National Institute of Environmental Research, Seo-gu, Incheon 22689, Korea; waterfa@korea.kr
5   Nanopharmaceutical Sciences, Kyung Hee University, Dongdaemun-gu, Seoul 02447, Korea
*   Correspondence: parkys@khu.ac.kr; Tel.: +82-2-961-0946

Received: 29 September 2017; Accepted: 14 November 2017; Published: 27 November 2017

Abstract: Aquatic oligochaetes are very common in streams, and are used as biological assessment indicators as well as in the biological management of organic-enriched systems. In this study, we analyzed the effects of environmental factors influencing the distribution of aquatic oligochaetes *Limnodrilus hoffmeisteri* in streams. We used 13 environmental factors in three categories (i.e., geography, hydrology, and physicochemistry). Data on the distribution of oligochaetes and environmental factors were obtained from 1159 sampling sites throughout Korea on a nationwide scale. Hierarchical cluster analysis (HCA) and nonmetric multidimensional scaling (NMDS) were performed to analyze the relationships between the occurrence of aquatic oligochaetes and environmental factors. A random forest model was used to evaluate the relative importance of the environmental factors affecting the distribution of oligochaetes. HCA classified sampling sites into four groups according to differences in environmental factors, and NMDS ordination reflected the differences of environmental factors, in particular, water depth, velocity, and altitude, among the four groups defined in the HCA. Furthermore, using a random forest model, turbidity and water velocity were evaluated as highly important factors influencing the distribution of *L. hoffmeisteri*.

Keywords: distribution patterns of species; environmental factor; multiple scale; multivariate analyses; machine learning model

---

1. Introduction

The distribution and abundance of organisms are governed by various environmental factors and, therefore, it is essential to understand the influence of environmental factors on ecological communities [1]. Freshwater oligochaetes are distributed in various freshwater habitats and have been widely applied as indicator species in environmental assessment [2–4]. The influences of hydrology and physicochemistry on stream oligochaetes have been reported in many studies [5–8]. For example, the distribution of oligochaetes has been shown to be related to salinity, temperature, dissolved oxygen (DO), current velocities, and habitat stability [9] and *Limnodrilus hoffmeisteri* has been observed to burrow deeper into sediment when exposed to hypoxic water [10]. Van Duinen et al. [11] assessed the effects of increased nutrient availability on aquatic oligochaetes, and Martinez-Ansemil and Collado [12] demonstrated that substrate type and current velocity are the principal variables explaining community structure. Other researchers reported that substrate type is important for oligochaete communities [13,14]. Sauter and Güde [14] presented that tubificids are more abundant in fine substrate. The proportions of functional feeding strategies did not follow the gradient of hydraulic conditions, and only a few oligochaete taxa were able to survive in hydraulically rough conditions, and most oligochaete taxa were found only in pools [15].

Their ecological functions and roles have been recognized in various studies, including bioprocessing ability for restoration of organic polluted freshwater habitats [16–19], sludge removal [20], improvement of water quality [21,22], behavioral changes in response to pollutants [23], sediment toxicity assessment [24–26], and response to several toxic substances [3,27,28]. In a recent literature review, Kang et al. [29] reported that three species (*Lumbriculus variegatus*, *Tubifex tubifex*, and *Limnodrilus hoffmeisteri*) among freshwater oligochaetes have been used most frequently in ecotoxicological studies, indicating the possibility of developing an early warning system using aquatic oligochaetes to monitor aquatic ecosystem disturbance.

Although there have been many studies on the distribution of oligochaetes, most studies have been conducted on the taxonomy [30–32]. However, limited studies have been conducted on the relationships between the distribution of oligochaetes and their habitats. Ecological data have non-linear and complex properties within variables as well as among variables. Therefore, nonlinear approaches are recommended to characterize the relationships among variables in ecological data [33,34]. Recently, machine learning algorithms, which are robust to nonlinear and complex data, are commonly used in environmental sciences and ecological studies [35,36]. Among machine learning techniques, the usage of random forest (RF), an ensemble machine learning technique, has been increasing in field of ecology, including in ecohydrological distribution modelling [37], species distribution modelling [38,39], and the relation between freshwater organisms and environmental factors [40,41], because in the computation of RF, no a priori assumptions about linear or nonlinear relationships between predictors and response variables are needed [42]. Furthermore, there are lots of reports that RF produces more accurate classification compared with other machine learning techniques (e.g., classification trees) because RF is computed based on the combinations of many decision trees [43]. In this study, we investigated the distribution patterns that improve, like that of the aquatic oligochaete *L. hoffmeisteri* in Korean streams, and the influence of environmental factors on their distribution patterns by using multivariate analyses and a machine learning algorithm.

2. Materials and Methods

2.1. Ecological Data

The data on aquatic oligochaetes used in this study was provided by the Nationwide Aquatic Ecological Monitoring Program (NAEMP), conducted by the Ministry of Environment and the National Institute of Environmental Research, Korea. We used a dataset compiled from 1159 sites sampled from 2009 to 2016 on a nationwide scale in Korea (Figure 1). The sampling was conducted from the headwater to downstream on a nationwide scale to reflect various habitat conditions, as well as to evaluate the status of freshwater habitats nationally. The altitude in sampling sites ranged from 0 to 1166 m a.s.l., the distance from source was from 0.2 to 498.9 km, and the stream order is from 1 to 7.

Benthic macroinvertebrates, including aquatic oligochaetes, were collected using a Surber net (30 × 30 cm², mesh size 1 mm) with three replicates at each site, twice a year in spring and summer, based on the guidelines of the National Survey for Stream Ecosystem Health in South Korea [44]. Oligochaeta samples were sorted and preserved in 70% ethanol in the laboratory. Then, for coarse identification, sorted individuals were separated into several possible genus/species groups based on morphological characters (e.g., body seta, body shape, etc.). The specimens were sorted into the lowest level (mostly species level) following the identification keys [45,46]. Among oligochaetes, we studied the distribution patterns of *L. hoffmeisteri*, which is the dominant species in Korean streams, and the influence of environmental factors on their distribution patterns. According to the literature [31,47], we used the species name *L. hoffmeisteri* instead of *Limnodrilus gotoi*, which was used in the database of the NAEMP.

We used the following 13 environmental parameters grouped into three categories, namely, geography, hydrology, and physicochemistry: altitude and slope in geography; water depth, water width, and current velocity in hydrology; and dissolved oxygen (DO), biochemical oxygen demand (BOD), total nitrogen (TN), total phosphorus (TP), chlorophyll a (Chl-*a*), pH, conductivity, and turbidity in physicochemistry (Table 1). Geographical factors were extracted from a digital map using a geographic information system (ArcGIS®, ver.10.1, ESRI [48]). Hydrological and physicochemical factors were obtained from the NAEMP database [49], which were measured according to the NAEMP guidelines [46].

As the number of samples collected differed among sampling sites, the average value for each site was used in the study of environmental variables. Among the 1159 sampling sites, we used 1127 sites in the analyses owing to the unavailability of environmental data for the other 32 sites.

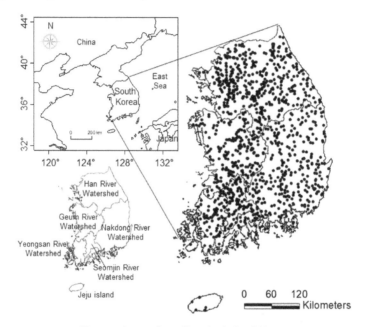

Figure 1. A map of sampling sites in South Korea.

Table 1. Differences (mean with standard error) in environmental factors among four different groups defined through a hierarchical cluster analysis. Different letters in the same row indicate significant differences between the groups based on Tukey's multiple comparison tests ($p < 0.05$). Dissolved oxygen (DO); biochemical oxygen demand (BOD); total nitrogen (TN); total phosphorus (TP); chlorophyll a (Chl-*a*).

Category	Variable	Group			
		1	2	3	4
Geography	Altitude (m)	183.64 (9.80) [a]	140.96 (9.83) [b]	69.12 (4.10) [d]	99.69 (5.37) [c]
	Slope (°)	2.76 (0.35) [a]	1.99 (0.40) [a]	0.59 (0.13) [b]	0.75 (0.11) [b]
Hydrology	Water width (m)	32.05 (1.89) [c]	30.04 (2.93) [c]	136.05 (11.12) [a]	54.25 (5.81) [b]
	Water depth (cm)	26.15 (0.54) [b]	22.97 (0.47) [c]	49.53 (1.15) [a]	25.51 (0.27) [b]
	Velocity (cm/s)	64.97 (0.75) [a]	37.38 (0.46) [b]	18.26 (0.83) [c]	14.95 (0.42) [d]
Physicochemistry	Biochemical oxygen demand (BOD) (mg/L)	1.62 (0.07) [c]	1.70 (0.08) [bc]	2.10 (0.07) [a]	1.89 (0.05) [ab]
	Total nitrogen (TN) (mg/L)	2.83 (0.09) [a]	2.52 (0.09) [bc]	2.67 (0.07) [ab]	2.40 (0.05) [c]
	Total phosphorus (TP) (mg/L)	0.10 (0.01)	0.10 (0.01)	0.10 (0.01)	0.08 (0.01)
	Chlorophyll a (Chl-*a*) (mg/L)	2.65 (0.19) [b]	3.17 (0.25) [b]	5.05 (0.38) [a]	3.26 (0.14) [b]
	Dissolved oxygen (DO) (mg/L)	9.29 (0.08) [a]	8.86 (0.08) [b]	8.74 (0.09) [b]	8.41 (0.06) [c]
	pH	7.77 (0.02) [b]	7.75 (0.03) [b]	7.88 (0.03) [a]	7.80 (0.02) [ab]
	Electric conductivity (μS/cm)	195.05 (10.08) [b]	239.53 (21.24) [b]	764.06 (202.54) [a]	285.68 (64.57) [b]
	Turbidity (Nephelometric Turbidity Units, NTU)	11.67 (0.77) [a]	7.74 (0.80) [b]	12.64 (0.60) [a]	13.02 (0.85) [a]

Note: Generally, alphabet "*a*" represent the highest value following [b], [c], [d], etc.

2.2. Data Analyses

Hierarchical cluster analysis (HCA) and nonmetric multidimensional scaling (NMDS) were performed to analyze the relationships between the occurrence (i.e., presence or absence) of aquatic oligochaetes and environmental factors at multiple spatial scales. Sampling sites were classified into several groups based on the similarities of their environmental conditions through a hierarchical HCA using the Ward linkage method with Euclidean distance measure. The same data used in HCA were used in NMDS, which is a method of scaling that places similarities between variables on a multidimensional scale [50]. In NMDS, a biplot was used to characterize the relationships among environmental factors, sampling sites, and occurrence of oligochaetes. Environmental factors such as water width, altitude, Chl-*a*, conductivity, and turbidity were log-transformed prior to analyses to reduce the variation in each variable. One-way analysis of variance (ANOVA) was performed to evaluate differences in the occurrence frequencies of oligochaetes and environmental factors among the groups defined in the HCA. Tukey's multiple comparison tests were performed if there were differences among groups in HCA.

A random forest (RF) model was used to evaluate the relative importance of environmental factors affecting the distribution of oligochaetes. After building the RF model with the 13 selected environmental factors as independent variables, and presence or absence of oligochaetes as a dependent variable, a measure of the total decrease in node impurity (IncNodePurity) was used to evaluate the relative importance of the 13 environmental factors. We evaluated the relationships between two of the most important factors and the occurrence of oligochaetes on a three-dimensional graph fitting a smooth surface to the points of oligochaete occurrence.

All analyses were performed in R [51]. HCA, ANOVA, and Tukey's test were conducted using the stats package [51] in R. NMDS was carried out using the metaMDS function of the vegan package [52], RF was performed using the randomForest package [53], and generation of three-dimensional graphs with a fitting surface model was conducted with the car package [54] in R.

3. Results

The sampling sites were classified into four groups according to similarities in their environmental factors through HCA (Figures 2 and 3). The percentage occurrence of *L. hoffmeisteri* in each group was in the range of 82.7% (in group 1) to 96.0% (in group 4) (Table 1). The groups had significantly different environmental characteristics. The sampling sites in groups 1 and 2 were located in areas characterized by high altitude and slope, whereas those in groups 3 and 4 were in areas with low values of altitude and slope. Velocity was significantly different among the four groups (ANOVA, $p < 0.05$) with the highest value in group 1 (65.0 cm/s) and the lowest value in group 4 (14.9 cm/s). The water width of groups 3 and 4, which showed high occurrence frequencies of oligochaetes, were 136.1 and 54.2 m, respectively, whereas width was low in groups 1 and 2, at 32.0 and 30.0 m, respectively. Physicochemical factors reflected the differences in geological and hydrological factors among groups. Groups 1 and 2 displayed lower BOD, conductivity, and turbidity, whereas these factors were higher in groups 3 and 4, indicating the relatively polluted condition at sampling sites in groups 3 and 4. DO and TN showed the opposite trend. TP was relatively constant without any significant differences among the four groups.

Sampling sites in group 1 were mainly from the Han River (Figure 3). Sites in group 2 were mostly distributed near the mountain range, and groups 1 and 2 contained few samples from Nakdong River. Groups 3 and 4 showed high occurrence frequency of *L. hoffmeisteri*. Geographical distribution reveals that sampling sites in group 3 were polluted downstream of Nakdong River with high values of BOD, conductivity and turbidity. The sites in group 4 had the lowest flow rate and DO, high conductivity and turbidity.

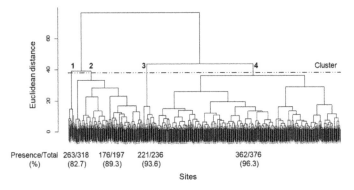

Figure 2. A dendrogram of hierarchical cluster analysis based on the similarities of 13 environmental factors. The dotted line indicates the clustering point. Presence/Total (%) indicates the number of sites with presence of *Limnodrilus hoffmeisteri* among the number of sites assigned in each group, and the percentage is the proportion of the number of sampling sites with presence of *L. hoffmeisteri*.

Figure 3. Distribution of sampling sites in each group defined from hierarchical cluster analysis. (a) group 1, (b) group 2, (c) group 3, and (d) group 4.

The results of NMDS (the stress in the first two axes was 0.034) analysis were consistent with those of HCA (Figure 4). Group 1 was arranged on the left side of the NMDS ordination, group 2 was in the center, group 3 was on the upper right, and group 4 was on the lower right. Axis 1 was highly correlated with altitude ($r = 0.80$, $p < 0.05$) in geography, current velocity ($r = 0.97$, $p < 0.05$) in hydrology, and DO ($r = 0.75$, $p < 0.05$) in physicochemistry, whereas axis 2 was highly correlated with water width ($r = 0.96$, $p < 0.05$) and water depth ($r = 0.97$, $p < 0.05$) in hydrology and turbidity ($r = 0.95$, $p < 0.05$) in physicochemistry. The occurrence of *L. hoffmeisteri* was plotted on the NMDS ordination of 13 environmental factors based on the correlation coefficient between coordinates of the NMDS ordination and occurrence of species, showing that the occurrence of *L. hoffmeisteri* characterized groups 3 and 4.

The RF model showed the high probability of predicting the presence or absence of *L. hoffmeisteri* with 1.0 of AUC (area under the curve), which is generally used to evaluate prediction models with presence or absence output. Therefore, we used this model to evaluate the importance of environmental variables affecting the distribution of oligochaetes. Turbidity, velocity, DO, and conductivity were accordingly identified as the most influential factors (Figure 5a). We evaluated the changes in oligochaete occurrence in response to changes in the two most important factors (the physicochemical factor turbidity and the hydrological factor velocity) based on a surface map using a smooth regression method (Figure 5b). As the velocity increased, the occurrence probability of oligochaetes decreased. In contrast, as turbidity increased, there was an increase in occurrence probability. Therefore, the highest occurrence probability of oligochaetes coincided with a low velocity and a high turbidity, whereas occurrence probability was low at very low turbidity with high velocity.

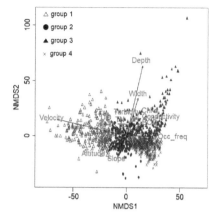

Figure 4. Nonmetric multidimensional scaling (NMDS) ordination of sampling sites based on the similarities of 13 environmental factors. The stress in the first two axes was 0.034. Occ_freq: occurrence frequency of *Limnodrilus hoffmeisteri*.

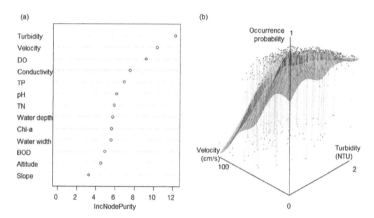

Figure 5. (a) The importance of environmental factors affecting the occurrence of *Limnodrilus hoffmeisteri* based on a random forest model. (b) Relationship between the two most important factors (water velocity and turbidity) and the occurrence of *L. hoffmeisteri* based on the random forest model.

4. Discussion

We investigated the distribution patterns of aquatic oligochaetes *L. hoffmeisteri* in relation to environmental factors in three categories. Sampling sites were classified according to altitude, and this geographical factor influenced hydrological and water quality factors. Our results showed that *L. hoffmeisteri* was mostly present at sites with low altitude and slope, wide water width, and a low velocity. In addition, on the basis of an RF model, current velocity and turbidity were identified as the most important factors influencing the distribution of *L. hoffmeisteri*. Therefore, local environmental conditions are important in determining the distribution of *L. hoffmeisteri* in our study even though large-scale environmental conditions (e.g., altitude) are also influential to its occurrence, supporting the environmental filtering theory for the distribution of species [55,56].

The habitat conditions for oligochaetes can be explained in terms of their ecological traits. Aquatic oligochaetes are negatively phototactic [57]. The rear end of the body burrows in the water, whereas the front of the body remains buried in the sediment [57,58]. Moreover, they are deposit-feeders, so they eat soft bottoms which are mainly present in the most lowland section of rivers, where water velocity is low with high turbidity and Chl-*a*. In addition, it is well known that oligochaetes can resist to oxygen depletion [59] and sediment contamination. Therefore, they can survive at sites with high anthropogenic contamination (e.g., high turbidity, high organic matter and presence of toxic substances).

The particle size of substrates is assumed to play a major role in the distribution and abundance of tubificids [16]. Generally, aquatic oligochaetes prefer silt clay, which has the smallest sediment particle size [60], and tubificid species in particular have preference for a particle size of less than 63 μm [61,62]. In a study on the hydrological and substrate determinants of oligochaete distribution, Verdonschot [7] showed that hydro-morphological parameters explained differences in the major distribution patterns.

On the basis of the aforementioned properties, aquatic oligochaetes, including *L. hoffmeisteri*, are used for bioprocessing ability in the restoration of organic-polluted freshwater habitats [16–19], and for the improvement of water quality [21,22]. Tubificid species are useful for the denitrification of NO_3^- [18] and the reduction of sludge [16,19] in the overlying water. In terms of water chemistry, major changes in the structure of oligochaete assemblages have been shown to be related to water mineralization. Martínez-Ansemil and Collado [12] reported that substrate and water velocity are the most important factors influencing the distribution of oligochaetes, but that water mineralization is also positively related to oligochaete distribution. Marchand [63] has also reported that DO and organic matter affect the distribution of oligochaetes. Organic matter and dissolved minerals are components of turbidity [64], and whereas oligochaetes are affected by water mineralization, these worms in turn also have an influence on water mineralization [18].

L. hoffmeisteri is one of the cosmopolitan oligochaetes [47] because it has tolerance to both organic and inorganic pollution, and exhibits variation in life history (duration of breeding period) characteristics according to the gradients of variously scaled environmental conditions [65,66]. Furthermore, unlike other tubificids, *L. hoffmeisteri* does not have a preference for silt and clay substrates [16], which accordingly contributes to this species' widespread distribution and abundance [65].

5. Conclusions

L. hoffmeisteri is distributed throughout South Korea, widely applied as biological indicators and a highly recommended candidate to mitigate organic-enriched freshwater ecosystem. However, up to now there has been limited studies on their ecological traits and favorable environmental conditions. In our study, we found out that its distribution is mainly determined by local conditions such as turbidity and water velocity based on the results of HCA, NMDS and RF. Our study, based on a nationwide scale database, showed that other local environmental factors such as water quality are also important. However, further studies including local to large scaled surveyed dataset are necessary to figure out its ecology and ecological traits deeply.

Acknowledgments: This work was supported by the National Research Foundation of Korea (NRF) funded by the Korean government (MSIP) (grant number NRF-2016R1A2B4011801). The data were collected under the project of "National Aquatic Ecosystem Health Survey and Assessment" supported by the Ministry of Environment and the National Institute of Environmental Research, Korea. The authors are grateful to survey members involved in the project, and also thank the anonymous reviewers for their helpful comments in improving the scientific content of the manuscript.

Author Contributions: H.K. prepared the dataset, performed data analyses, and wrote the manuscript. M.-J.B. performed data analyses, and wrote the manuscript. S.-J.H., and J.-S.M. provided the dataset, and D.-S.L. assisted data processing. Y.-S.P. developed the concept of the study, performed data analysis and wrote the manuscript. All the authors contributed to the review of the manuscript.

Conflicts of Interest: The authors declare no conflict of interest. The founding sponsors had no role in the design of the study; in the collection, analyses, or interpretation of data; in the writing of the manuscript, and in the decision to publish the results.

References

1. Bae, M.J.; Li, F.; Kwon, Y.S.; Chung, N.; Choi, H.; Hwang, S.J.; Park, Y.S. Concordance of diatom, macroinvertebrate and fish assemblages in streams at nested spatial scales: Implications for ecological integrity. *Ecol. Indic.* **2014**, *47*, 89–101. [CrossRef]

2. Brinkhurst, R.O.; Kennedy, C.R. Studies on the biology of the Tubificidae (Annelida, Oligochaeta) in a polluted stream. *J. Anim. Ecol.* **1965**, *34*, 429–443. [CrossRef]

3. Chapman, P.M.; Farrell, M.A.; rinkhurst, R.O. Relative tolerances of selected aquatic oligochaetes to individual pollutants and environmental factors. *Aquat. Toxicol.* **1982**, *2*, 47–67. [CrossRef]

4. Lin, K.J.; Yo, S.P. The effect of organic pollution on the abundance and distribution of aquatic oligochaetes in an urban water basin, Taiwan. *Hydrobiolgia* **2008**, *596*, 213–223. [CrossRef]

5. Korn, H. Studien zur Ökologie der Oligochaeten in der oberen Donau unter Berücksichtigung der Abwasserwirkungen. *Archiv Für Hydrobiol.* **1963**, *27*, 131–182. (In German)

6. Prenda, J.; Gallardo, A. The influence of environmental factors and microhabitat availability on the distribution of an aquatic oligochaete assemblage in a Mediterranean river basin. *Int. Rev. Hydrobiol.* **1992**, *77*, 421–434. [CrossRef]

7. Verdonschot, P.F.M. Hydrology and substrates: Determinants of oligochaete distribution in lowland streams (The Netherlands). *Hydrobiologia* **2001**, *463*, 249–262. [CrossRef]

8. Schenková, J.; Helešic, J. Habitat preferences of aquatic oligochaeta (Annelida) in the Rokytná River, Czech Republic—A small highland stream. *Hydrobiologia* **2006**, *564*, 117–126. [CrossRef]

9. Pascar-Gluzman, C.; Dimentman, C. Distribution and habitat characteristics of Naididae and Tubificidae in the inland waters of Israel and the Sinai Peninsula. *Hydrobiologia* **1984**, *115*, 197–205. [CrossRef]

10. Fischer, J.A.; Beeton, A.M. The effect of dissolved oxygen on the burrowing behavior of *Limnodrilus hoffmeisteri* (Oligochaeta). *Hydrobiologia* **1975**, *47*, 273–290. [CrossRef]

11. Van Duinen, G.A.; Timm, T.; Smolders, A.J.; Brock, A.M.; Verberk, W.C.; Esselink, H. Differential response of aquatic oligochaete species to increased nutrient availability—A comparative study between Estonian and Dutch raised bogs. *Hydrobiologia* **2006**, *564*, 143–155. [CrossRef]

12. Martínez-Ansemil, E.; Collado, R. Distribution patterns of aquatic oligochaetes inhabiting watercourses in the Northwestern Iberian Peninsula. *Hydrobiologia* **1999**, *334*, 73–83. [CrossRef]

13. Dumnicka, E. Communities of oligochaetes in mountain streams of Poland. *Hydrobiologia* **1994**, *278*, 107–110. [CrossRef]

14. Sauter, G.; Güde, H. Influence of grain size on the distribution of tubificid oligochaete species. *Hydrobiologia* **1996**, *334*, 97–101. [CrossRef]

15. Syrovátka, V.; Schenkova, J.; Brabec, K. The distribution of chironomid larvae and oligochaetes within a stony-bottomed river stretch: The role of substrate and hydraulic characteristics. *Fundam. Appl. Limnol. Archiv Fur Hydrobiol.* **2009**, *174*, 43–62. [CrossRef]

16. Ratsak, C.H.; Verkuijlen, J. Sludge reduction by predatory activity of aquatic oligochaetes in waste water treatment plants: Science or fiction? A review. *Hydrobiologia* **2006**, *564*, 197–211. [CrossRef]

17. Mermillod-Blondin, F.; Nogaro, G.; Datry, T.; Malard, F.; Gibert, J. Do tubificid worms influence the fate of organic matter and pollutants in stormwater sediments? *Environ. Pollut.* **2005**, *134*, 57–69. [CrossRef] [PubMed]

18. Pelegrí, S.P.; Blackburn, T.H. Effects of *Tubifex tubifex* (Oligochaeta: Tubificidae) on N-mineralization in freshwater sediments, measured with 15N isotopes. *Aquat. Microb. Ecol.* **1995**, *9*, 289–294. [CrossRef]

19. Rensink, J.H.; Rulkens, W.H. Using metazoa to reduce sludge production. *Water Sci. Technol.* **1997**, *36*, 171–179.

20. Elissen, H.J.H.; Mulder, W.J.; Hendrickx, T.L.G.; Elbersen, H.W.; Beelen, B.; Temmink, H.; Buisman, C.J.N. Aquatic worms grown on biosolids: Biomass composition and potential applications. *Bioresour. Technol.* **2010**, *101*, 804–811. [CrossRef] [PubMed]

21. Choi, Y.H. The Blood Worm (*Limnodrilus socialis*) Using Capacity for Treatment of Aquaculture Wastewater. Master's Thesis, Chungju University, Chungju-si, Korea, 2005.

22. Jun, T.S.; Park, J.H. The blood worm, *Limnodrilus socialis*'s using capacity for treatment of aquaculture wastewater. *Chungju Univ. Theses Collect.* **2005**, *40*, 201–206.

23. Macedo-Sousa, J.; Gerhardt, A.; Brett, C.M.A.; Nogueira, A.; Soares, A.M.V.M. Behavioural responses of indigenous benthic invertebrates (*Echinogammarus meridionalis*, *Hydropsyche pellucidula* and *Choroterpes picteti*) to a pulse of acid mine drainage: A laboratorial study. *Environ. Pollut.* **2008**, *156*, 966–973. [CrossRef] [PubMed]

24. Dermott, R.; Munawar, M. A simple and sensitive assay for evaluation of sediment toxicity using *Lumbriculus variegatus* (Muller). In *Sediment/Water Interactions*; Springer: Dordrecht, The Netherlands, 1992; pp. 407–414.

25. Phipps, G.L.; Ankley, G.T.; Benoit, D.A. Use of the aquatic oligochaeta *Lumbriculus variegatus* for assessing the toxicity and bioaccumulation of sediment-associated contaminants. *Environ. Toxicol. Chem.* **1993**, *12*, 269–279. [CrossRef]

26. Ingersoll, C.G.; Brunson, E.L.; Wang, N.; Dwyer, F.J.; Ankley, G.T.; Mount, D.R.; Huckins, J.N.; Petty, J.; Landrum, P.F. Uptake and depuration of nonionic organic contaminants from sediment by the oligochaete, *Lumbriculus variegatus*. *Environ. Toxicol. Chem.* **2003**, *22*, 872–885. [CrossRef] [PubMed]

27. Lucan-Bouché, M.L.; Biagianti-Risbourg, S.; Arsac, F.; Vernet, G. An original decontamination process developed by the aquatic oligochaete *Tubifex tubifex* exposed to copper and lead. *Aquat. Toxicol.* **1999**, *45*, 9–17. [CrossRef]

28. Rathore, R.S.; Khangarot, B.S. Effects of temperature on the sensitivity of sludge worm *Tubifex tubifex* Müller to selected heavy metals. *Ecotoxicol. Environ. Saf.* **2002**, *53*, 27–36. [CrossRef] [PubMed]

29. Kang, H.; Bae, M.-J.; Park, Y.-S. Ecotoxicological studies using aquatic oligochaetes: Review. *Korean J. Ecol. Environ.* **2016**, *49*, 343–353. [CrossRef]

30. Lee, J.; Jung, J. Four unrecorded species of tubificid oligochaetes (Annelida: Clitellata) in Korea. *Anim. Syst. Evol. Divers.* **2014**, *30*, 240–247. [CrossRef]

31. Park, H.J.; Timm, T.; Bae, Y.J. Aquatic oligochaete (Annelida: Clitellata) fauna from the Jungnang Stream in Seoul, Korea, with eight new Korean records. *Korean J. Ecol. Environ.* **2013**, *46*, 507–512.

32. Park, H.J.; Timm, T.; Bae, Y.J. Taxonomy of the Korean freshwater Oligochaeta (Annelida) with eight species new to Korea. *Entomol. Res. Bull.* **2013**, *29*, 180–188.

33. Park, Y.-S.; Chang, J.; Lek, S.; Cao, W.; Brosse, S. Conservation strategies for endemic fish species threatened by the Three Gorges Dam. *Conserv. Biol.* **2003**, *17*, 1748–1758. [CrossRef]

34. Park, Y.-S.; Tison, J.; Lek, S.; Coste, M.; Giraudel, J.; Delmas, F. Application of a self-organizing map in ecological informatics: Selection of representative species from large community dataset. *Ecol. Inf.* **2006**, *1*, 247–257. [CrossRef]

35. Park, Y.-S.; Céréghino, R.; Compin, A.; Lek, S. Applications of artificial neural networks for patterning and predicting aquatic insect species richness in running waters. *Ecol. Model.* **2003**, *160*, 265–280. [CrossRef]

36. Park, Y.-S.; Kwon, Y.-S.; Hwang, S.-J.; Park, S.-K. Characterizing effects of landscape and morphometric factors on water quality of reservoirs using a Self-organizing map. *Environ. Model. Softw.* **2014**, *55*, 214–221. [CrossRef]

37. Peters, J.; De Baets, B.; Verhoest, N.E.; Samson, R.; Degroeve, S.; De Becker, P.; Huybrechts, W. Random forests as a tool for ecohydrological distribution modelling. *Ecol. Model.* **2007**, *207*, 304–318. [CrossRef]

38. Araújo, M.B.; New, M. Ensemble forecasting of species distributions. *Trends Ecol. Evol.* **2007**, *22*, 42–47. [CrossRef] [PubMed]

39. Thomaes, A.; Kervyn, T.; Maes, D. Applying species distribution modelling for the conservation of the threatened saproxylic Stag Beetle (*Lucanus cervus*). *Biol. Conserv.* **2008**, *141*, 1400–1410. [CrossRef]

40. Bae, M.J.; Kwon, Y.; Hwang, S.J.; Chon, T.S.; Yang, H.J.; Kwak, I.S.; Park, J.H.; Ham, S.A.; Park, Y.S. Relationships between three major stream assemblages and their environmental factors in multiple spatial scales. *Ann. Limnol.-Int. J. Limnol.* **2011**, *47*, S91–S105. [CrossRef]

41. Kwon, Y.S.; Li, F.; Chung, N.; Bae, M.J.; Hwang, S.J.; Byoen, M.S.; Park, S.J.; Park, Y.S. Response of fish communities to various environmental variables across multiple spatial scales. *Int. J. Environ. Res. Public Health* **2012**, *9*, 3629–3653. [CrossRef] [PubMed]

42. Prasad, A.M.; Iverson, L.R.; Liaw, A. Newer classification and regression tree techniques: Bagging and random forests for ecological prediction. *Ecosystems* **2006**, *9*, 181–199. [CrossRef]

43. Cutler, D.R.; Edwards, T.C.; Beard, K.H.; Cutler, A.; Hess, K.T.; Gibson, J.; Lawler, J.J. Random forests for classification in ecology. *Ecology* **2007**, *88*, 2783–2792. [CrossRef] [PubMed]

44. Ministry of Environment (MOE); National Institute of Environmental Research (NIER). *The Survey and Evaluation of Aquatic Ecosystem Health in Korea*; NIER: Incheon, Korea, 2008; pp. 43–72.

45. Brinkhurst, R.O.; Jamieson, B.G.M. *Aquatic Oligochaeta of the World*; Oliver and Body: Edinburgh, Scotland, 1971.

46. Brinkhust, R.O. *Guide to the Freshwater Aquatic Microdrile Oligochaetes of North America*; Canadian Special Publication of Fisheries and Aquatic Sciences: Ottawa, ON, Canada, 1986; Volume 84, p. 259.

47. Timm, T. Distribution of freshwater oligochaetes in the west and east coastal regions of the North Pacific Ocean. *Hydrobiologia* **1999**, *406*, 67–81. [CrossRef]

48. Environmental Systems Research Incorporated (ESRI). *ArcGIS 10.1*; Environmental Systems Research Incorporated: Redlands, CA, USA, 2012. Available online: https://www.esri.com (accessed on 26 November 2017).

49. Water Resource Management Information System (WAMIS). Seoul, Republic of Korea. Available online: http://www.wamis.go.kr (accessed on 3 March 2017).

50. Legendre, P.; Legendre, L.F. *Numerical Ecology*, 3rd ed.; Elsevier: Oxford, UK, 2012; pp. 508–520. ISBN 9780444538680.

51. R Core Team. *R: A Language and Environment for Statistical Computing*; R Foundation for Statistical Computing: Vienna, Austria, 2017; ISBN 3-900051-07-0. Available online: https://www.r-project.org/ (accessed on 3 March 2017).

52. Oksanen, J.; Blanchet, F.G.; Friendly, M.; Kindt, R.; Legendre, P.; McGlinn, D.; Minchin, P.R.; O'Hara, R.B.; Simpson, G.L.; Solymos, P.; et al. Vegan: Community Ecology Package; R Package Version 2.4-4. 2017. Available online: https://CRAN.R-project.org/package=vegan (accessed on 24 April 2017).

53. Liaw, A.; Wiener, M. Classification and regression by randomForest. *R News* **2002**, *2*, 18–22. Available online: https://cran.r-project.org/web/packages/randomForest/ (accessed on 7 June 2017).

54. Fox, J.; Weisberg, S. *An {R} Companion to Applied Regression*, 2nd ed.; SAGE: Thousand Oaks, CA, USA, 2011. Available online: https://cran.r-project.org/web/packages/car/index.html (accessed on 3 March 2017).

55. Poff, N.L. Landscape filters and species traits: Towards mechanistic understanding and prediction in stream ecology. *J. N. Am. Benthol. Soc.* **1997**, *16*, 391–409. [CrossRef]

56. Lamouroux, N.; Dolédec, S.; Gayraud, S. Biological traits of stream macroinvertebrate communities: Effects of microhabitat, reach, and basin filters. *J. N. Am. Benthol. Soc.* **2004**, *23*, 449–466. [CrossRef]

57. Timm, T.; Martin, P.J. Clitellata: Oligochaeta. In *Freshwater Invertebrates*, 4th ed.; Thorp, J., Rogers, D.C., Eds.; Academic Press: Boston, MA, USA, 2015; pp. 529–549. ISBN 9780123850263.

58. Avel, M.A.R.C.E.L. Classe des annélides oligochètes. *Trait. De Zool.* **1959**, *5*, 224–470. (In French)

59. Håkanson, L. The relationship between salinity, suspended particulate matter and water clarity in aquatic systems. *Ecol. Res.* **2006**, *21*, 75–90. [CrossRef]

60. Lazim, M.N.; Learner, M.A. The influence of sediment composition and leaf litter on the distribution of tubificid worms (Oligochaeta). *Oecologia* **1987**, *72*, 131–136. [CrossRef] [PubMed]

61. Rodriguez, P.; Martinez-Madrid, M.; Arrate, J.A.; Navarro, E. Selective feeding by the aquatic oligochaete *Tubifex tubifex* (Tubificidae, Clitellata). *Hydrobiologia* **2001**, *463*, 133–140. [CrossRef]

62. Kang, H.; Bae, M.-J.; Park, Y.-S. Behavioral response of *Tubifex tubifex* to changes of water temperature and substrate composition. *Korean J. Ecol. Environ.* **2017**, *50*, 355–361.

63. Marchand, J. The influence of seasonal salinity and turbidity maximum variations on the nursery function of the Loire estuary (France). *Aquat. Ecol.* **1993**, *27*, 427–436. [CrossRef]

64. Bilotta, G.S.; Brazier, R.E. Understanding the influence of suspended solids on water quality and aquatic biota. *Water Res.* **2008**, *42*, 2849–2861. [CrossRef] [PubMed]

65. Kennedy, C.R. The life history of *Limnodrilus hoffmeisteri* Clap. (Oligochaeta: Tubificidae) and its adaptive significance. *Oikos* **1966**, *17*, 158–168. [CrossRef]

66. Poddubnaya, T.L. Life cycles of mass species of Tubificidae (Oligochaeta). In *Aquatic Oligochaete Biology*; Springer: Boston, MA, USA, 1980; pp. 175–184.

Article

Diversity and Distribution of Endemic Stream Insects on a Nationwide Scale, South Korea: Conservation Perspectives

Mi-Jung Bae [1] and Young-Seuk Park [2],*

[1] Freshwater Biodiversity Research Division, Nakdonggang National Institute of Biological Resources, Sangju-si, Gyeongsangbuk-do 37242, Korea; mjbae@nnibr.re.kr

[2] Department of Life and Nanopharmaceutical Sciences, and Department of Biology, Kyung Hee University, Seoul 02447, Korea

* Correspondence: parkys@khu.ac.kr; Tel.: +82-02-961-0946

Received: 30 September 2017; Accepted: 24 October 2017; Published: 30 October 2017

Abstract: This study aimed to identify the biogeographical and environmental factors affecting the biodiversity of endemic aquatic species (i.e., Ephemeroptera, Plecoptera and Trichoptera; EPT). We used data collected from 714 sampling sites combined with 39 environmental factors. Ten EPT endemic species were identified. The sampling sites grouped into four clusters based on the similarities of the endemic EPT assemblages using a hierarchical cluster analysis. Non-metric multidimensional scaling (NMS) revealed the differences among the four clusters, with the first three axes being strongly related to annual average, August, and January temperatures, as well as altitude. The random forest model identified geological and meteorological factors as the main factors influencing species distribution, even though the contributions of environmental factors were species-specific. Species with the lower occurrence frequency (i.e., *Pteronarcys macra*, *Kamimuria coreana*, and *Psilotreta locumtenens*) mainly occurred in the least-disturbed habitats. *P. macra* represents a priority conservation species, because it has a limited distribution range and is highly vulnerable to anthropogenic disturbance. Our results support the need for an environmental management policy to regulate deforestation and conserve biodiversity, including endemic species.

Keywords: aquatic insects; endemic species; EPT taxa; species conservation; random forest model; nonmetric multidimensional scaling; stream ecosystem; biodiversity

---

1. Introduction

Endemic species generally inhabit a geologically limited area, and are highly vulnerable to small environmental changes [1]. Therefore, we must strive to conserve and manage such endemic species from loss and extinction. However, knowledge remains limited on the biodiversity of endemic species in freshwater ecosystems, along with the key factors that influence their distribution. Severe pressures from various anthropogenic disturbances are continuously causing changes to original habitats and threatening the continued existence of endemic species [1,2].

Freshwater habitats cover just 0.8% of the Earth's surface and comprise around 0.01% of the world's water bodies [3,4]. However, freshwater habitats contain disproportionally high biodiversity of organisms [5]. Freshwater organisms tend to inhabit smaller geographic ranges [6], resulting in high levels of endemism evolving. Severe anthropogenic disturbances have been further shrinking their potential distribution ranges, as well as individual home ranges. Dudgeon et al. [7] listed five major factors that threaten freshwater biodiversity: overexploitation (e.g., [8]); water pollution (e.g., [9]); flow modification (e.g., [10]), and; the destruction and degradation of habitat and invasion of exotic species (e.g., [11]). The impacts of these factors, both separately and in combination, have caused the populations of freshwater organisms to decline, along with their distributions, and the homogenization of freshwater ecosystem worldwide. These factors threaten the survival of endemic species, which are highly adapted to the environmental factors within their specific ranges [12].

Research on the endemism, diversity, distribution, and conservation efforts of endemic species remains limited, especially with respect to endemic macroinvertebrates at nationwide scales. Most literature on endemic species focuses on single species (i.e., *Lednia tumana*, [13]). Other studies have been conducted within specific areas or under specific environmental conditions, such as with respect to the effects of glacial melting [13], deforestation [14], and the introduction of invasive plants [15].

Among various ecologically similar species, rare species tend to have smaller populations that are more likely to be vulnerable to environmental events, resulting in a greater extinction risk than common species [16]. Therefore, the degree of rarity of endemic species should be the first criterion used to determine conservation priorities and conservation responsibilities [17]. Among aquatic insects, species in Ephemeroptera, Plecoptera, and Trichoptera (EPT) exhibit sensitive responses to physical environmental factors at broad scales, in addition to water quality factors at small scales. The diversity of EPT represents one of the most important biological indices for evaluating the status of freshwater habitats [18]. However, few studies have focused on the rarity and distribution of EPT endemic species [19], despite their high vulnerability to future environmental change [1,20–23] and their ecological impacts and roles in ecosystem functioning [24]. To our knowledge, even though several studies have investigated the endemism of EPT species, they mainly focus on how climate change or temperature related factors influence the vulnerability of ETP species [13,19].

In this study, we evaluated two hypotheses: first, endemic species diversity becomes higher in the least disturbed areas, and; second, large scale factors, such as meteorological and geological factors (e.g., temperature, latitude, and longitude) are important determinants of the distribution and occurrence of endemic species, even though habitat preference and environmental tolerance differ among endemic species. Our results are expected to provide baseline information on which to build suggestions to advance the conservation efforts of endemic species.

2. Materials and Methods

2.1. Ecological Data

Data on endemic EPT species were obtained from the database of the National Aquatic Ecological Monitoring Program (NAEMAP), supported by the Ministry of Environment and National Institute of Environmental Research, Republic of Korea [25]. From the NAEMAP database, we selected a dataset consisting of samples collected at 717 sampling sites of 371 streams in five major river basins at a national scale in South Korea from 2008 to 2013. South Korea has a temperate climate with an annual mean temperature of 12.8 °C and an annual mean precipitation of 1589 mm/year [26]. The Han River basin (basin area: 41,957 km^2) runs through Seoul, the capital of South Korea, and is the largest basin in South Korea (covering approximately one third of the country). The basin is located in the northern part of South Korea, and flows westward into the Yellow Sea. The Nakdong River basin (31,785 km^2) is located in the southeastern part of South Korea, and flows into the South Sea. The Guem River basin (17,537 km^2) is in the mid-western part of South Korea and the Yeongsan River basin (12,833 km^2) is in the southwestern part of the country. Both basins flow into the Yellow Sea. The Seomjin River basin (4914 km^2) is located between the Nakdong and Yeongsan river basins, and flows into the South Sea. The Yeongsan and Seomjin rivers were treated as one river system (hereafter, the Yeongseom River) for management purposes, because their catchments are in quite close proximity and share similar geographical conditions [23].

At each sampling site, three replicate sampling surveys were conducted in riffle zones within a 200 m reach of the river using a Surber net (30 × 30 cm^2, 1 mm mesh size), based on the guidelines of the National Survey for Stream Ecosystem Health in South Korea [27]. All specimens collected during sampling were sorted and identified to the lowest level (mostly species level) under the microscope based on the published literature [28–32]. We used 39 environmental factors from seven different categories (including meteorology, geography, land use, hydrology, flow type, substrate, and physicochemical water quality) to evaluate how these factors affect EPT endemic species. The factors of hydrology, flow type, substrate, and physicochemical water quality were obtained from the NAEMAP database [33]. Meteorological factors were obtained from the Korea Meteorological Administration. Geographical and land use factors were extracted from digital maps in ArcGIS (Ver. 10.1) [34] based on the coordinate information of each sampling site [25].

2.2. Data Analysis

We conducted data analyses in four steps to characterize the rarity, abundance, and distribution of endemic EPT species. First, we evaluated the rarity of endemic EPT species by considering the number of sampling sites where endemic EPT species were found and the number of endemic EPT species recorded at each individual sampling site at nationwide and basin scales. Second, a hierarchical cluster analysis (CA) based on the endemic EPT assemblage was conducted to define the pattern of similarity of the endemic EPT assemblage structure. CA was calculated based on the Ward's linkage method [35] with the Bray-Curtis distance measure. Then, a Kruskal Wallis test (KW) was conducted to compare the differences of endemic EPT assemblage structure among the clusters defined in the CA. When there were significant differences in KW ($p < 0.05$), multiple comparison tests were performed to compare the differences between clusters. Multi-response permutation procedures (MRPP) were conducted to check for significant differences among the clusters. CA and MRPP were conducted with a function hclust and mrpp, respectively, in the package vegan [36] in R [37]. KW and multiple comparison tests were carried out with the function kruskal in the package agricolae [38] in R [37].

Third, non-metric multidimensional scaling (NMS) was conducted using the Bray-Curtis distance as the dissimilarity measure for the endemic EPT assemblage to determine the distribution pattern of endemic EPT species. NMS is a non-linear method that is suitable for zero-inflated ecological data sets with unknown data distribution [39]. We used a metaMDS function to determine the best solution (i.e., the lowest stress value) in the package vegan [36] in R. The relationship between endemic EPT species and environmental factors was determined using a function envfit in the package vegan [39,40].

Finally, the occurrence probability of endemic species was predicted using a random forest (RF) model, using the 39 environmental factors as independent variables [41]. RF is an ensemble machine learning technique that is based on a combination of a large set in decision tree. Each tree is trained to select a random sample (i.e., the calibration data set) from a random set and the training dataset of variables [42]. After building the RF model, the relative importance of environmental factors influencing the occurrence of endemic species was calculated from a mean decrease in accuracy, and was then rescaled from 0 to 100. In this study, RF was computed using the package random Forest [43] in R with three training parameters (such as mtree, mtry, and node size) at the default setting. The abundance of each species and some of the environmental factors with high variation (i.e., distance from source, water width, average depth, and average velocity) were log transformed before the data analysis.

3. Results

3.1. Characteristics of Endemic EPT Species

Ten endemic EPT species were identified in the dataset, including five, three, and two species belonging to Ephemeroptera, Plecoptera, and Trichoptera, respectively (Table 1). Out of these 10 species, *Rhoenanthus coreanus* (Ephemeroptera) was the most widely distributed, with 42.2% occurrence frequency (301 sites), and was found within all four basins (Figure 1 and Table 1). However, most species had very low occurrence frequencies. *Pteronarcys macra* (Plecoptera) was only recorded at eight sites (1.1% of the occurrence frequency) in only one basin (the Han River basin) in the northern part of South Korea (Figure 1, Table 1). Only four species (i.e., *R. coreanus, Drunella aculea, Potamanthus yooni,* and *Kamimuria coreana*) had more than 10% occurrence frequency. Out of these four species, only two (*R. coreanus* and *K. coreana*) were observed in all four basins, even though the occurrence frequency of *K. coreana* was extremely low in the Geum and Yeongseom basins (3.1% in both cases). In addition, out of six species with less than 10.0% of the occurrence frequency, only one (*Neoperla coreensis*) was recorded in all basins. Overall, the majority of sites did not have any endemic species (i.e., 277 sites, 38.6%) or just one endemic species (i.e., 219 sites, 30.5%) (Figure 2).

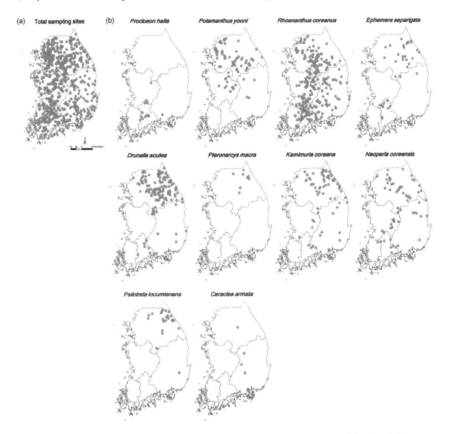

Figure 1. Sampling sites (**a**) and occurrence patterns of endemic EPT species (**b**) in South Korea.

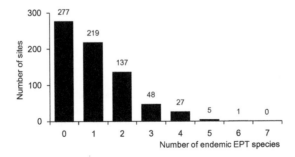

Figure 2. Number of sampling sites for endemic EPT species.

Table 1. Occurrence frequency of endemic Ephemeroptera, Plecoptera and Trichoptera (EPT) species recorded in South Korea from 2008 to 2013.

Order	Species	Abbreviation	River Basin (%)				
			Total	Han	Nakdong	Guem	Yeongseum
Ephemeroptera	*Procloeon halla*	Proc	14 (2.0)	0 (0.0)	1 (0.8)	1 (0.8)	12 (9.0)
	Potamanthus yooni	Pota	86 (12.0)	67 (20.9)	4 (3.1)	15 (11.5)	0 (0.0)
	Rhoenanthus coreanus	Rhoe	301 (42.2)	112 (35.0)	57 (43.8)	59 (45.4)	73 (54.5)
	Ephemera separigata	Ephe	36 (5.0)	23 (7.2)	2 (1.5)	0 (0.0)	11 (8.2)
	Drunella aculea	Drun	151 (21.1)	135 (42.2)	10 (7.7)	0 (0.0)	6 (4.5)
Plecoptera	*Pteronarcys macra*	Pter	8 (1.1)	8 (2.5)	0 (0)	0 (0.0)	0 (0.0)
	Kamimuria coreana	Kami	75 (10.5)	56 (17.5)	11 (8.5)	4 (3.1)	4 (.03)
	Neoperla coreensis	Neop	69 (9.7)	33 (10.3)	13 (10.0)	14 (10.8)	9 (6.7)
Trichoptera	*Psilotreta locumtenens*	Psil	25 (3.5)	24 (7.5)	1 (0.8)	0 (0.0)	0 (0.0)
	Ceraclea armata	Cera	11 (1.5)	1 (0.3)	9 (6.9)	0 (0.0)	1 (0.0)

3.2. Relationships between Endemic EPT Species and Environmental Factors

The sampling sites were classified into four clusters (1 to 4) based on the similarities of the endemic EPT assemblage composition (Figure 3). MRPP identified significant differences in the endemic EPT assemblage among the four clusters (A = 0.09, $p < 0.05$). Cluster 1 was characterized by *D. aculea*, *P. macra*, *K. coreana*, *N. coreensis*, *Psilotreta locumtenens*, and *Ceraclea armata* (KW, $p < 0.05$, Table 2). Cluster 2 was characterized by *R. coreanus*. Cluster 3 was characterized by *P. yooni*. Cluster 4 was characterized by *Procloeon halla*.

Differences in the composition of the endemic EPT assemblage were also reflected in NMS (Figure 4). In NMS ordination, the first three axes (stress value = 7.32) had the highest relationship with annual average temperature ($R^2 = 0.598$, $p < 0.05$), followed by August temperature ($R^2 = 0.561$, $p < 0.05$), January temperature ($R^2 = 0.503$, $p < 0.05$), latitude ($R^2 = 0.445$, $p < 0.05$), and altitude ($R^2 = 0.373$, $p < 0.05$) (Table 3). The sampling sites with high values for latitude, water velocity, and altitude were located on the left parts of axis 1, whereas sites with relatively high values for (Chl-a), poor water quality (i.e., high conductivity, biological oxygen demand (BOD), total phosphate (TP), and chlorophyll a (Chl-a)) were on the right part of axis 1 (Figure 4). Species such as *P. halla*, were located on the right part of axis 1. These species mainly inhabit the lowland areas of the southern parts of South Korea, where the sampling areas were mostly characterized as having high conductivity, BOD, TP, and Chl-a. Species, such as *P. locumtenens*, *K. coreana*, *P. macra*, and *D. aculea*, were located on the left part of axis 1. These species were mainly found in the least disturbed freshwater habitats (e.g., with good water quality, high water velocity, and altitude), such as mountain areas.

Table 2. Differences in the abundance of endemic EPT species among four clusters from a cluster analysis. The values in parenthesis are the standard deviation. Different alphabets indicate significant differences of variables among clusters based on the multiple comparison tests ($p < 0.05$).

Order	Species	Cluster			
		1	2	3	4
Ephemeroptera	*Procloeon halla*	0.00 (0.00) [b]	0.03 (0.10) [a]	0.00 (0.00) [b]	0.32 (2.26) [a]
	Potamanthus yooni	0.42 (2.52) [b]	0.16 (0.49) [b]	8.22 (15.28) [a]	0.00 (0.00) [c]
	Rhoenanthus coreanus	0.46 (1.23) [c]	25.24 (39.77) [a]	9.47 (19.77) [b]	0.88 (0.68) [b]
	Ephemera separigata	0.13 (0.48) [a]	0.09 (0.52) [ab]	0.02 (0.11) [b]	0.04 (0.16) [ab]
	Drunella aculea	11.83 (16.3) [a]	0.06 (0.32) [b]	0.03 (0.13) [b]	0.04 (0.14) [b]
Plecoptera	*Pteronarcys macra*	0.35 (2.62) [a]	0.00 (0.00) [b]	0.00 (0.00) [b]	0.00 (0.00) [b]
	Kamimuria coreana	2.51 (6.35) [a]	0.12 (0.65) [b]	0.01 (0.05) [b]	0.02 (0.08) [b]
	Neoperla coreensis	0.19 (0.63) [a]	1.23 (3.56) [a]	0.02 (0.09) [b]	0.03 (0.12) [b]
Trichoptera	*Psilotreta locumtenens*	1.13 (4.57) [a]	0.00 (0.00) [b]	0.00 (0.00) [b]	0.00 (0.00) [b]
	Ceraclea armata	0.09 (0.56) [a]	0.00 (0.04) [b]	0.00 (0.00) [b]	0.00 (0.04) [ab]

Table 3. Relationship between environmental factors and NMS ordination based on endemic EPT assemblages. The environmental factors of the 10 highest R^2 are presented as boldface letters.

Environmental Factor	Abbreviation	NMS1	NMS2	NMS3	R^2	p
Geology						
Latitude	Lati_X	**−0.81**	**0.56**	**0.20**	**0.45**	**0.001**
Longitude	Longi_Y_	−0.89	−0.46	−0.06	0.24	0.001
Altitude (m)		**−0.85**	**−0.08**	**0.51**	**0.37**	**0.001**
Slope (°)		−0.82	−0.43	0.39	0.19	0.001
Distance from source	DFS	0.85	0.45	0.29	0.08	0.001
Meteorological factors						
Annual average temperature (°C)	Ave_temp	**0.93**	**−0.15**	**−0.34**	**0.60**	**0.001**
August temperature (°C)	Aug_temp	**0.95**	**0.09**	**−0.29**	**0.56**	**0.001**
January temperature (°C)	Jan_temp	**0.84**	**−0.37**	**−0.40**	**0.50**	**0.001**
Thermal range (°C)	Thermal_range	−0.20	0.91	0.37	0.20	0.001
Precipitation (mm)		−0.09	−0.90	−0.42	0.05	0.002
Land use (%)						
Urban		0.36	0.47	−0.81	0.10	0.001
Agriculture		0.86	0.47	0.20	0.09	0.001
Forest		−0.78	−0.54	0.33	0.20	0.001
Grass		0.48	0.25	−0.84	0.01	0.281
Bareland		0.98	0.19	0.04	0.02	0.044
Hydrology						
Stream order	Str_order	0.87	0.46	0.20	0.23	0.001
Water width (m)	W_width	0.84	0.53	0.09	0.12	0.001
Average depth (cm)	Ave_depth	0.62	0.75	0.24	0.12	0.001
Average velocity (cm/s)	Ave_velocity	**−0.93**	**0.26**	**0.25**	**0.35**	**0.001**
Flow type						
Riffle		−0.95	−0.26	0.19	0.29	0.001
Run		0.84	0.42	−0.35	0.08	0.001
Pool		1.00	−0.02	0.10	0.07	0.001
Substrate (%)						
Silt		0.67	0.31	−0.68	0.02	0.037
Clay		0.66	0.70	−0.28	0.10	0.001
Sand		0.44	0.68	−0.59	0.22	0.001
Small pebble		0.84	−0.54	−0.04	0.04	0.003
Pebble		−0.12	−0.80	0.59	0.04	0.002
Cobble		−0.61	−0.46	0.64	0.22	0.001
Boulder		−0.73	−0.63	0.28	0.18	0.001
Physicochemical water quality						
Biological oxygen demand (mg/L)	BOD	**0.78**	**0.57**	**−0.27**	**0.33**	**0.001**
Total nitrogen (mg/L)	TN	−0.03	1.00	−0.09	0.15	0.001
Ammonia nitrogen (mg/L)	NH3N	0.55	0.82	−0.16	0.05	0.002
Nitrate nitrogen (mg/L)	NO3N	0.23	0.97	0.12	0.13	0.001
Total phosphate (mg/L)	TP.	**0.83**	**0.49**	**−0.25**	**0.31**	**0.001**
Phosphate-phosphorus (mg/L)	PO4P	0.91	0.40	−0.12	0.21	0.001
Chlorophyll a (mg/L)	Chl.a	**0.88**	**0.43**	**−0.22**	**0.38**	**0.001**
Dissolved oxygen (mg/L)	DO	−0.68	0.58	0.46	0.19	0.001
pH	pH	0.80	0.46	0.39	0.10	0.001
Conductivity (um/S)	Conductivity	**0.73**	**0.63**	**−0.269**	**0.32**	**0.001**

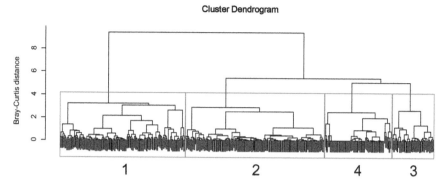

Figure 3. Dendrogram of cluster analysis based on the endemic EPT assemblages.

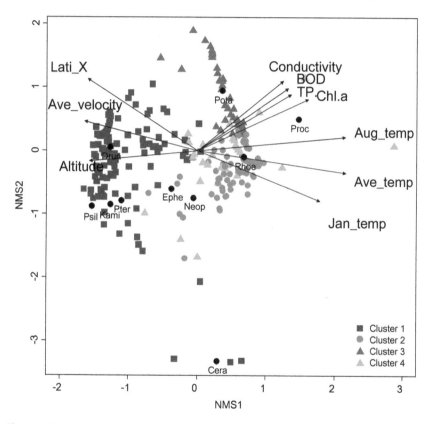

Figure 4. Non-metric multidimensional scaling (NMS) ordination based on endemic EPT assemblages with fitted vectors of environmental factors. Black circles with four letters represent species, and others without letters indicate sampling sites. The direction and length of arrows represent the strength of the relationship between the environmental variables and the ordination axes. Only 10 environmental factors with R^2 values > 0.3 are displayed in the figure. Abbreviation s for species and environmental factors are presented in Tables 1 and 3, respectively. Cluster 1: square with green color; cluster 2: circle with red color; cluster 3: triangle with blue color, and; cluster 4: triangle with orange color.

3.3. Influential Environmental Factors on the Occurrences of Endemic Species

The distribution of each endemic species was predicted well in the range from 0.990 to 0.998 based on environmental factors through the RF learning process (Figure 5). Overall, geographical and meteorological factors represented the main factors influencing species distribution, even though different species responded differently to various environmental factors. For example, altitude was the most important (100) for the occurrence of *R. coreanus*, which had the highest occurrence frequency in the dataset, followed by pebbles (90.4), longitude (83.1), distance from source (77.4), and annual average temperature (64.7). Meanwhile, *D. aculea*, which was the second highest occurrence frequency species, was characterized by January temperature (100) and annual average temperature (95.9). *P. macra*, which rarely occurred, was influenced by the ratio of agriculture area in land use (100), followed by the run ratio in flow type (86.8), forest ratio (85.7), phosphate-phosphorus (PO4P) (84.0), and average depth (52.0).

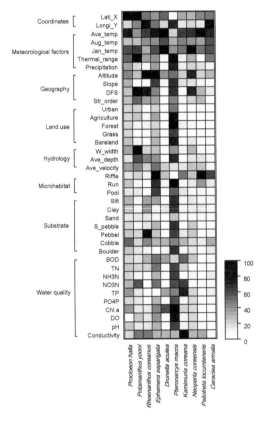

Figure 5. Relative importance in the occurrence of endemic EPT species based on the Random Forest model. Abbreviations of environmental factors are presented in Table 3.

4. Discussion

4.1. Priority Species for the Conservation

Endemic species are characterized by their limited spatial distribution and poor dispersal, resulting in their being rare. The rarity of endemic species is a major causal factor of their going extinct in both ecological and geological timeframes [44]. Therefore, endemic species are likely to be the first candidates for extinction. In our study, out of the 10 endemic EPT species, the occurrence frequency of six species was less than 10.0%. The frequency of *C. armata* (1.5%) and *P. macra* (1.1%) was lower than 2.0%. The species that were widely distributed across several basins might lose their conservation priority when compared with species that are only found within a certain basin and with a limited distribution [12]. In this sense, both species should have high conservation priority. In particular, *P. macra* should be considered as priority species of conservation concern, because it was only found in a limited northeastern area of the Han River, which has a low annual temperature. *P. macra* is a cold-adapted species, making it vulnerable to global warming and anthropogenic disturbances, which threaten its existence in their original habitat [23].

4.2. Influential Environmental Factors Related to the Existence of Endemic Species: Conservation Implications

We should be aware of increasing threats to the existence of endemic species. Not surprisingly, endemism is an important criterion for determining national conservation responsibilities [17]. However, measurements for the future extinction of species are hindered by the lack knowledge of the life history, traits, niche, resource requirements, and suitable criteria of endemic species making it difficult to determine their rarity [45,46]. The current study showed that the factors influencing the differentiation of endemic EPT species included geology (i.e., latitude and altitude), meteorology (especially temperature-related factors), hydrology (water velocity), and water chemistry (BOD, TP, Chl-a, and conductivity).

Our results support the environmental filtering hypothesis, which proposes that environmental drivers act as hierarchical filters constraining assemblages [47]. Large-scale factors, such as geological and meteorological factors, strongly influence the local habitat and biological diversity of streams and rivers [48,49]. These factors are closely connected with the diversity and composition of endemic EPT species [6]. Strayer [6] suggested that some range boundaries are set by the climate and other ecological patterns, even though the current distribution patterns of endemic EPT species is based on the history of drainage connections [50]. Freshwater fauna are particularly sensitive and vulnerable to the impacts of climate change, because they usually have limited dispersal abilities [51]. Consequently, current and future changes caused from human activities are likely to have a stronger influence on their diversity than past anthropogenic alterations [52]. Global warming especially threats the potential future distribution and persistence of the sensitive habitats used by endemic Plecoptera [13,53], even though this phenomenon was not directly considered in our study. Several studies have also shown that land use is an influential factor in determining species distribution, even though, in our study, this influence was relatively weak based on NMS. Flather et al. [54] found that forests and rangelands are important factors for differentiating endangered species "hot spots."

When the suitable habitats and their climatic refuges are degraded, the existence of endemic taxa becomes threatened (e.g., [55]). Since the 20th century, intensive alterations of hydrology have been closely related with the massive constructions of reservoirs and dams, as well as the disappearance of streams and springs, due to human actions [56]. Dams and impoundments alter the hydrological regimes of rivers, resulting in reduced water flow (water level fluctuations), the accumulation of silt, and the loss of habitat diversity. These alterations induce changes of life cycles, block of species dispersal, and reduce the abundance of freshwater fauna [57–59]. If these threats align within the zone containing high endemic species richness and if there are no conservation actions in that zone, it would accelerate the further homogenization of these habitats and the simplification of the macroinvertebrate community, which, in turn, would precipitate the loss of rare endemic species.

In addition, we found that endemic species with the lower occurrence frequency (i.e., *P. macra*, *K. coreana* and *P. locumtenens*) were mainly observed in the least-disturbed habitats, such as mountain areas (northerneast part of the Han River catchment). Certain factors (such as low temperature, high altitude, high ratio of forest land use, rapid water velocity, high ratio of riffles, and good water quality) strongly influenced the distribution of these three species based on RF. These three species primarily inhabit zones with little urbanization and extensive forest watersheds. However, the forest area in

South Korea has gradually declined (i.e., 6.406×10^6 ha in 2003 and 6.335×10^6 ha in 2015) because of the construction of convenience facilities for humans. Furthermore, the ratio of private forests is high (4.25×10^6 ha, 67.1% of the total forest area); thus, such forest areas are likely to be quickly lost due to potentially disruptive activities by the owners. Therefore, a strict environmental management policy is required to minimize the deforestation of such areas to conserve biodiversity, including endemic species.

5. Conclusions

The conservation, protection, and management of freshwater biodiversity are the ultimate conservation challenge because multiple human stakeholders continue to threaten biodiversity. Various efforts to maintain and protect biodiversity should be conducted globally through establishing a system of protected areas. We suggest that endemic invertebrates should be priority candidate species of conservation concern based on their rarity and the types of environmental factors that determine their diversity and distribution. Out of the 10 species that were found as endemic species in South Korea, *P. macra* had an occurrence frequency of just 1.1% with an extremely limited distribution (i.e., found in one catchment, especially the least disturbed area), indicating the high conservation priority of this species. RF showed that the distribution of the endemic species was mainly influenced by geographical and meteorological factors. Benthic macroinvertebrates play an important role in many freshwater ecosystems, and are useful for evaluating biological integrity and water and habitat quality. Furthermore, the high diversity of endemic EPT species is a valuable predictor of the diversity of other aquatic invertebrates. Therefore, conservation efforts of sites containing endemic EPT species also guarantee the conservation of other freshwater taxa.

Acknowledgments: We thank the National Aquatic Ecological Monitoring Program (NAEMP) operated by the Ministry of Environment and National Institute of Environmental Research, Korea for providing part of the dataset used. This work was supported by the Nakdonggang National Institute of Biological Resources (Project: Studies on Freshwater Biodiversity in Headwater Streams of the Nakdong River Basin) and by the National Research Foundation of Korea (NRF) funded by the Korean government (MSIP) (grant number NRF-2016R1A2B4011801).

Author Contributions: M.-J.B. and Y.-S.P. conceived the study. M.-J.B. conducted the statistical analysis and interpretation of the data. M.-J.B. and Y.-S.P. wrote and revised the manuscript.

Conflicts of Interest: The authors declare no conflict of interest.

References

1. Brown, B.L. Habitat heterogeneity and disturbance influence patterns of community temporal variability in a small temperate stream. *Hydrobiologia* **2007**, *586*, 93–106. [CrossRef]
2. Winterbourn, M.J.; Cadbury, S.; Ilg, C.; Milner, A.M. Mayfly production in a New Zealand glacial stream and the potential effect of climate change. *Hydrobiologia* **2008**, *603*, 211–219. [CrossRef]
3. Gleick, P.H. Water resources. In *Encyclopedia of Climate and Weather*; Schneider, S.H., Ed.; Oxford University Press: Oxford, UK, 1996; pp. 817–823.
4. Geist, J. Integrative freshwater ecology and biodiversity conservation. *Ecol. Indic.* **2011**, *11*, 1507–1516. [CrossRef]
5. Hawksworth, D.L.; Kalin-Arroyo, M.T. Magnitude and distribution of biodiversity. In *Global Biodiversity Assessment*; Heywood, V.H., Watson, R.T., Eds.; Cambridge University Press: Cambridge, UK, 1995; pp. 107–192.
6. Strayer, D.L. Challenges for freshwater invertebrate conservation. *J. N. Am. Benthol. Soc.* **2006**, *25*, 271–287. [CrossRef]
7. Dudgeon, D.; Arthington, A.H.; Gessner, M.O.; Kawabata, Z.-I.; Knowler, D.J.; Lévêque, C.; Naiman, R.J.; Prieur-Richard, A.-H.; Soto, D.; Stiassny, M.L.J.; et al. Freshwater biodiversity: Importance, threats, status and conservation challenges. *Biol. Rev.* **2006**, *81*, 163–182. [CrossRef] [PubMed]
8. Cowx, I.G.; Collares-Pereira, M.J.; Coelho, M.M. *Freshwater Fish Conservation: Options for the Future*; Fishing News: Oxford, UK, 2002; pp. 443–452.
9. Xu, M.; Wang, Z.; Duan, X.; Pan, B. Effects of pollution on macroinvertebrates and water quality bio-assessment. *Hydrobiologia* **2014**, *729*, 247–259. [CrossRef]

10. Nilsson, C.; Reidy, C.A.; Dynesius, M.; Revenga, C. Fragmentation and flow regulation of the world's large river systems. *Science* **2005**, *308*, 405–408. [CrossRef] [PubMed]

11. Koehn, J.D. Carp (*Cyprinus carpio*) as a powerful invader in Australian waterways. *Freshw. Biol.* **2004**, *49*, 882–894. [CrossRef]

12. Burlakova, L.E.; Karatayev, A.Y.; Karatayev, V.A.; May, M.E.; Bennett, D.L.; Cook, M.J. Endemic species: Contribution to community uniqueness, effect of habitat alteration, and conservation priorities. *Biol. Conserv.* **2011**, *144*, 155–165. [CrossRef]

13. Muhlfeld, C.C.; Giersch, J.J.; Hauer, F.R.; Pederson, G.T.; Luikart, G.; Peterson, D.P.; Downs, C.C.; Fagre, D.B. Climate change links fate of glaciers and an endemic alpine invertebrate. *Clim. Chang.* **2011**, *106*, 337–345. [CrossRef]

14. Benstead, J.P.; Pringle, C.M. Deforestation alters the resource base and biomass of endemic stream insects in eastern Madagascar. *Freshw. Biol.* **2004**, *49*, 490–501. [CrossRef]

15. Samways, M.J.; Sharratt, N.J.; Simaika, J.P. Effect of alien riparian vegetation and its removal on a highly endemic river macroinvertebrate community. *Biol. Invasions* **2011**, *13*, 1305–1324. [CrossRef]

16. Flather, C.H.; Sieg, C.H. Species rarity: Definition, causes and classification. In *Conservation of Rare or Little-Known Species*; Island Press: Washington, DC, USA, 2007; pp. 40–66.

17. Schmeller, D.S.; Gruber, B.; Budrys, E.; Framsted, E.; Lengyel, S.; Henle, K. National responsibilities in European species conservation: A methodological review. *Conserv. Biol.* **2008**, *22*, 593–601. [CrossRef] [PubMed]

18. Couceiro, S.R.M.; Hamada, N.; Forsberg, B.R.; Pimentel, T.P.; Luz, S.L.B. A macroinvertebrate multimetric index to evaluate the biological condition of streams in the Central Amazon region of Brazil. *Ecol. Indic.* **2012**, *18*, 118–125. [CrossRef]

19. Brown, L.E.; Céréghino, R.; Compin, A. Endemic freshwater invertebrates from southern France: Diversity, distribution and conservation implications. *Biol. Conserv.* **2009**, *142*, 2613–2619. [CrossRef]

20. Céréghino, R.; Lavandier, P. Influence of hypolimnetic hydropeaking on the distribution and population dynamics of Ephemeroptera in a mountain stream. *Freshw. Biol.* **1998**, *40*, 385–399. [CrossRef]

21. Santoul, F.; Figuerola, J.; Mastrorillo, S.; Céréghino, R. Patterns of rare fish and aquatic insects in a southwestern French river catchment in relation to simple physical variables. *Ecography* **2005**, *28*, 307–314. [CrossRef]

22. Li, F.; Chung, N.; Bae, M.J.; Kwon, Y.S.; Park, Y.S. Relationships between stream macroinvertebrates and environmental variables at multiple spatial scales. *Freshw. Biol.* **2012**, *57*, 2107–2124. [CrossRef]

23. Li, F.; Chung, N.; Bae, M.J.; Kwon, Y.S.; Kwon, T.S.; Park, Y.S. Temperature change and macroinvertebrate biodiversity: Assessments of organism vulnerability and potential distributions. *Clim. Chang.* **2013**, *119*, 421–434. [CrossRef]

24. Wagenhoff, A.; Townsend, C.R.; Matthaei, C.D. Macroinvertebrate responses along broad stressor gradients of deposited fine sediment and dissolved nutrients: A stream mesocosm experiment. *J. Appl. Ecol.* **2012**, *49*, 892–902. [CrossRef]

25. Bae, M.J.; Kwon, Y.; Hwang, S.J.; Chon, T.S.; Yang, H.J.; Kwak, I.S.; Park, J.H.; Ham, S.A.; Park, Y.S. Relationships between three major stream assemblages and their environmental factors in multiple spatial scales. *Ann. Limnol. Int. J. Limnol.* **2011**, *47*, S91–S105. [CrossRef]

26. Lee, K.S.; Wenner, D.B.; Lee, I. Using H-and O-isotopic data for estimating the relative contributions of rainy and dry season precipitation to groundwater: Example from Cheju Island, Korea. *J. Hydrol.* **1999**, *222*, 65–74. [CrossRef]

27. MOE; NIER. *The Survey and Evaluation of Aquatic Ecosystem Health in Korea*; The Ministry of Environment and National Institute of Environmental Research: Incheon, Korea, 2008. (In Korean)

28. Quigley, M. *Invertebrates of Streams and Rivers: A Key to Identification*; Edward Arnold: London, UK, 1977.

29. Pennak, W. *Freshwater Invertebrates of the United States*; John Wiley and Sons, Inc.: New York, NY, USA, 1978.

30. Brighnam, A.R.; Brighnam, W.U.; Gnika, A. *Aquatic Insects and Oligochaeta of North and South Carolina*; Midwest Aquatic Enterprise: Mahomet, IL, USA, 1982.

31. Yun, I.B. *Illustrated Encyclopedia of Fauna and Flora of Korea. Aquatic Insects*; Ministry of Education: Seoul, Korea, 1988.

32. Merritt, R.W.; Cummins, K.W. *An Introduction to the Aquatic Insects of North America*; Hunt Publishing Company: Dubugue, IA, USA, 2006.

33. WAMIS. Available online: http://www.wamis.go.kr (accessed on 21 August 2014).

34. Environmental Systems Research Incorporated (ESRI). ArcGIS 10.0. Environmental Systems Research Incorporated: Redlands, CA, USA, 2011.

35. Ward JHJ. Hierarchical grouping to optimize an objective function. *J. Am. Stat. Assoc.* **1963**, *58*, 236–244.

36. Oksanen, J.; Kindt, R.; Legendre, P.; O'Hara, B.; Simpson, G.L.; Stevens, M.H.H.; Wagner, H. Vegan: Community Ecology Package. R Package Version 1.15-0. Available online: http://vegan.r-forge.r-project.org (accessed on 26 October 2017).

37. R Development Core Team. *R: A Language and Environment for Statistical Computing, 3-900051-07-0*; R Foundation for Statistical Computing: Vienna, Austria, 2011; Available online: http://www.R-project.org (accessed on 26 October 2017).

38. De Mendiburu, F. Agricolae: Statistical Procedures for Agricultural Research R Package Version 1.0-9. Available online: http://CRAN.R-project.org/package=agricolae (accessed on 1 April 2017).

39. Minchin, P.R. An evaluation of the relative robustness of techniques for ecological ordination. *Vegetatio* **1987**, *69*, 89–107. [CrossRef]

40. Virtanen, R.; Ilmonen, J.; Paasivirta, L.; Muotka, T. Community concordance between bryophyte and insect assemblages in boreal springs: A broad-scale study in isolated habitats. *Freshw. Biol.* **2007**, *54*, 1651–1662. [CrossRef]

41. Breiman, L. Random forests. *Mach. Learn.* **2001**, *45*, 5–32. [CrossRef]

42. Vincenzi, S.; Zucchetta, M.; Franzoi, P.; Pellizzato, M.; Pranovi, F.; De Leo, G.A.; Torricelli, P. Application of a Random Forest algorithm to predict spatial distribution of the potential yield of *Ruditapes philippinarum* in the Venice lagoon, Italy. *Ecol. Model.* **2011**, *222*, 1471–1478. [CrossRef]

43. Liaw, A.; Wiener, M. Classification and regression by random Forest. *R News* **2002**, *2*, 18–22.

44. Dobson, F.S.; Yu, J.; Smith, A.T. The importance of evaluating rarity. *Conserv. Biol.* **1995**, *9*, 1648–1651. [CrossRef]

45. Kuussaari, M.; Bommarco, R.; Heikkinen, R.K.; Helm, A.; Krauss, J.; Lindborg, R.; Ockinger, E.; Pärtel, M.; Pino, J.; Rodà, F.; et al. Extinction debt: A challenge for biodiversity conservation. *Trends Ecol. Evol.* **2009**, *24*, 564–571. [CrossRef] [PubMed]

46. Pimm, S.L.; Russell, G.J.; Gittleman, J.L.; Brooks, T.M. The future of biodiversity. *Science* **1995**, *269*, 347–350. [CrossRef] [PubMed]

47. De Bello, F.; Lavorel, S.; Lavergne, S.; Albert, C.H.; Boulangeat, I.; Mazel, F.; Thuiller, W. Hierarchical effects of environmental filters on the functional structure of plant communities: A case study in the French Alps. *Ecography* **2013**, *36*, 393–402. [CrossRef]

48. Allan, J.D. Landscapes and riverscapes: The influence of land use on stream ecosystems. *Annu. Rev. Ecol. Syst.* **2004**, *35*, 257–284. [CrossRef]

49. Allan, J.D.; Flecker, A.S. Biodiversity conservation in running waters. *BioScience* **1993**, *43*, 32–43. [CrossRef]

50. Van der Schalie, H. The value of mussel distribution in tracing stream confluence. *Mich. Acad. Sci. Arts Lett.* **1945**, *30*, 355–373.

51. Woodward, G.; Perkins, D.M.; Brown, L.E. Climate change and freshwater ecosystems: Impacts across multiple levels of organization. *Philos. Trans. R. Soc. B* **2010**, *365*, 2093–2106. [CrossRef] [PubMed]

52. Döll, P.; Zhang, J. Impact of climate change on freshwater ecosystems: A global-scale analysis of ecologically relevant river flow alterations. *Hydrol. Earth Syst. Sci.* **2010**, *14*, 783–799. [CrossRef]

53. Li, F.; de Figueroa, J.M.T.; Lek, S.; Park, Y.S. Continental drift and climate change drive instability in insect assemblages. *Sci. Rep.* **2015**, *5*, 11343. [CrossRef] [PubMed]

54. Flather, C.H.; Knowles, M.S.; Kendall, I.A. Threatened and endangered species geography. *BioScience* **1988**, *48*, 365–376. [CrossRef]

55. Deacon, J.E.; Williams, A.E.; Williams, C.D.; Williams, J.E. Fueling population growth in Las Vegas: How large-scale groundwater withdrawal could burn regional biodiversity. *BioScience* **2007**, *57*, 688–698. [CrossRef]

56. Søndergaard, M.; Jeppesen, E. Anthropogenic impacts on lake and stream ecosystems, and approaches to restoration. *J. Appl. Ecol.* **2007**, *44*, 1089–1094. [CrossRef]

57. Petts, G.E. *Impounded Rivers: Perspectives for Ecological Management*; Wiley: Chichester, UK, 1984; p. 285.

58. Vaughn, C.C.; Taylor, C.M. Impoundments and the decline of freshwater mussels: A case study of an extinction gradient. *Conserv. Biol.* **1999**, *13*, 912–920. [CrossRef]
59. Watters, G.T. Freshwater mussels and water quality: A review of the effects of hydrologic and instream habitat alterations. In *Freshwater Mollusk Symposium Proceedings*; Tankersley, R.A., Warmolts, D.I., Watters, G.T., Armitage, B.J., Johnson, P.D., Butler, R.S., Eds.; Ohio Biological Survey: Columbus, OH, USA, 2000; pp. 261–274.

Review

Evolutionary Toxicology as a Tool to Assess the Ecotoxicological Risk in Freshwater Ecosystems

Marianna Rusconi [1], Roberta Bettinetti [2], Stefano Polesello [1] and Fabrizio Stefani [1,*]

[1] Water Research Institute, National Research Council (IRSA-CNR), via Mulino 19, 20861 Brugherio, Italy; rusconi@irsa.cnr.it (M.R.); polesello@irsa.cnr.it (S.P.)

[2] Department of Theoretical and Applied Sciences, University of Insubria, Via Valleggio 11, 22100 Como, Italy; roberta.bettinetti@uninsubria.it

* Correspondence: stefani@irsa.cnr.it; Tel.: +39-039-2169-4232

Received: 17 October 2017; Accepted: 11 April 2018; Published: 17 April 2018

Abstract: Borrowing the approaches of population genetics, evolutionary toxicology was particularly useful in assessing the transgenerational effects of a substance at sublethal concentrations, as well as evaluating genetic variation in populations exposed to pollutants. Starting from assays in controlled conditions, in recent years this approach has also found successful applications multi-stressed natural systems. It is also able to exploit the huge amount of data provided by Next Generation Sequencing (NGS) techniques. Similarly, the focus has shifted from effects on the overall genetic variability, the so-called "genetic erosion", to selective effects induced by contaminants at more specific pathways. In the aquatic context, effects are usually assessed on non-model species, preferably native fish or macroinvertebrates. Here we provide a review of current trends in this specific discipline, with a focus on population genetics and genomics approaches. In addition, we demonstrate the potential usefulness of predictive simulation and Bayesian techniques. A focused collection of field and laboratory studies is discussed to demonstrate the effectiveness of this approach, covering a range of molecular markers, different endpoints of genetic variation, and different classes of chemical contaminants. Moreover, guidelines for a future implementation of evolutionary perspective into Ecological Risk Assessment are provided.

Keywords: genetic variability; adaptation; freshwater pollution; risk assessment

1. Introduction

In recent years, many genetic approaches have been adopted to investigate the effects of contaminants on gene transcription and evolutionary change. Among these approaches, gene expression (either surveying a set of traditional candidate genes or undertaking a genome-wide analysis) has been widely applied [1], whereas the potential of quantitative genetics and population genetics have only been partially explored [2,3].

In this context, evolutionary toxicology, despite being described more than 20 years ago [4], has only recently been proposed for ecotoxicological assessment [5–8]. In its original definition, the term borrows from the approach of population genetics and aims to identify causal links between toxicant pressures and changes in genetic variability at the population level [9] by applying the approaches and indicators that have proven to be informative in sister disciplines, such as conservation genetics [10,11].

Under this vision, evolutionary toxicology not only fills a gap in the scale of ecotoxicological assessment, but also introduces a novel vision to estimate long-term extinction risk with the awareness that contemporary evolutionary changes are common, widespread, and detectable after only a few generations of contaminant exposure. Evolutionary toxicology, taking into account multigenerational effects, can reveal the effects of substances (or mixtures of substances) at sublethal environmental concentrations and over long-term, chronic exposure, both of which are high-priority concerns in making an ecological risk assessment. In fact, effects can accumulate over generations and lead to reductions in fitness and, potentially, to population extinction. In one of the milestone manuscripts of evolutionary toxicology [7], four main research areas were inferred from the existing literature (i.e., the four cornerstones). Contaminants were divided in two main categories: those with a direct effect on nucleic acids (genotoxicants) and those with indirect effects on genetic variability mediated by fitness traits (non-genotoxicants) (Figure 1). Basically, non-genotoxic contaminants may reduce reproductive success, which can increase genetic drift and drive genome-wide changes in genetic diversity, which is referred to as genetic erosion. Additionally, toxicant effects on specific targets or fitness-related traits may determine changes in the allelic or genotypic frequencies as a result of selection at survivorship loci, not involving genome-wide changes but eventually producing side effects, such as increased susceptibility to other contaminants or to other environmental perturbations (fitness trade-offs). Finally, non-genotoxic substances may have cascade effects on behavioral traits, e.g., by limiting or altering the pattern of gene flow among populations. In this case, inbreeding or genetic drift may increase in an isolated population, leading to a loss of gene diversity and creating further fitness-related issues.

Genotoxic substances may have a direct effect on the genome by increasing the mutation rate and altering the balance between purifying the selection and the onset of novel alleles. This effect can be more significant in the case of mildly deleterious alleles, which are less severely purged by selection. The accumulation of deleterious mutations over generations in a population may lead to a mutational load: the negative additive effect of many slightly disadvantageous alleles on fitness traits [9].

Just six years after Bickham's pillar paper [7], Brady et al. [12], introducing a special issue on evolutionary toxicology, provided a paradigmatic analysis of recent trends in this discipline, which clearly set up more ambitious aims and targets thanks to the routine application of Next Generation Sequencing (NGS) techniques in this field. Indeed, the different research areas of evolutionary toxicology have recently coalesced into the characterization of evolutionary patterns, processes, and consequences of adaptation to toxicant exposure (Figure 1). This approach has benefited from the comparative investigation of exposed and control populations of potentially any species on the whole-genome scale as well as from quantitative genetics and epigenetics. In this context, top-down approaches, which aim at identifying the mechanisms underlying chemical resistance or the genomic regions impacted without any a priori knowledge, become prevalent [13]. Moreover, a strong tendency toward the integration of genomic analysis with different experimental approaches, including transcriptomics, proteomics, metabolomics, quantitative genetics, and environmental chemistry, was clearly demonstrated. The intention is to maximally exploit the potential of the NGS data, but more importantly, to better describe the relationships between the exposure of populations to contaminants and their long-term risk of impairment, which is the basis for an effective risk assessment. This task now appears less challenging than just a decade ago because, for example, the direct or indirect fitness costs of adaptation can now be demonstrated by integrating an analysis of phenotypic plasticity (e.g., through transcriptomics scans) with the genomic characterization of exposed and reference populations [14].

Regrettably, evolutionary changes can be evaluated only at a multigenerational scale, limiting their application as a routine effects-based method, as reviewed and discussed in a recent European Guideline [15]. Nevertheless, it is now a common scientific view that evolutionary toxicology will have a role in filling a gap that traditionally exists in ecotoxicological assessments, namely the scarcity of approaches for assessing population effects. By employing population genetic studies, evolutionary toxicology can potentially fill a void that exists between traditional ecotoxicity testing on individual organisms and the assessment of chemical effects on community structure and function.

This paper aims to evaluate the strengths and limitations of evolutionary toxicology as a tool to assess the ecotoxicological risk in freshwater ecosystems by reviewing the existing case studies and referring to novel analytical and statistical approaches that have been tested and developed in molecular ecology.

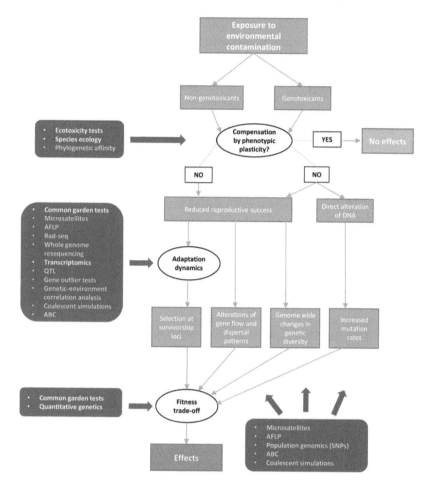

Figure 1. The four main cornerstones of evolutionary toxicology [7], derived from the exposure of populations to non-genotoxicant and genotoxicant contaminants, are presented in light blue. The white blocks represent the main processes leading to the four major responses on an evolutionary scale, which are relevant in the framework of an Ecological Risk Assessment procedure. In the dark blue blocks, the main approaches available to investigate these processes are listed. The techniques specifically discussed in this review are presented in bold yellow. The relevance of contamination for a population may first be evaluated by comparing the sensitivity of the target species with the severity of the expected exposure, and considering the phylogenetic affinity to model species and the potential for physiological compensation. The dynamics of adaptation, genetic erosion, or behavioral alterations may be revealed by a wide range of molecular techniques, including population genetics and genomics, and advanced statistical tools, such as approximate Bayesian computation (ABC). Finally, effects at an evolutionary scale, expressed as fitness trade-offs, can be demonstrated by exposing contaminated populations to singular or multiple stressors.

2. The Implementation of Population Genetics and Genomics in Evolutionary Toxicology: Markers, Techniques, and Data Analysis

The use of genetic techniques to evaluate chemical impacts is relatively recent, and the number of studies in the literature that have applied these techniques to the assessment of freshwater systems is limited, although it is continually growing. Since the first papers conceptualized this discipline [6,16], no more than a few dozen studies specifically addressing the issue of evolutionary toxicology have been published. Moreover, it is only since 2011, with the publication of a special issue of the journal Ecotoxicology on evolutionary toxicology (Reference [5], and papers within the same issue), that this term has been cited and used as keyword. Because a single review with a complete and in-depth discussion of all available approaches to evolutionary responses (e.g., population genetics and genomics, transcriptomics, quantitative genetics, quantitative trait loci, genome-wide association studies, epigenetics) is not possible, we focus primarily on population genetics and genomics, considering the integration with the other approaches where possible. We feel that because evolutionary toxicology is a young discipline, there is still the need to discuss, adapt, and transfer experiences and practices from sister disciplines, such as population genetics and genomics, to this specific field of study. We also consider population genetics and genomics to be one of, if not the most, relevant approaches to address evolutionary responses to contaminants. Specific reviews dealing with the integration of other approaches to ecotoxicology are available [2,17–20].

Signals of genetic alterations (i.e., reduced genetic variability) congruent with toxicant-induced stress were more frequently found in aquatic ecosystems than in terrestrial systems, probably because gene flows are subject to lower interference in aquatic drainages than in open spaces [21]. To date, fishes and macroinvertebrates have been preferably selected as model organisms. Generally, as in traditional toxicological tests, the ecology of the model organisms and their tolerance to toxicants, at least in terms of acute toxicity, should be well characterized and at least a partially sequenced genome should be available, particularly when applying genomic approaches. The molecular markers used for evolutionary toxicology surveys were those typically employed in traditional population genetic studies. Among them, microsatellites have been widely applied. Microsatellite loci are not expected to be under direct selection, although they might be linked to loci that are involved in resistance mechanisms [22–24]. More likely, these markers sample across the genome in a random way, making them useful for measuring the effects of pressures on overall genetic diversity. For organisms with an unknown genome, other versatile techniques such as Amplified Fragment Length Polymorphism (AFLP) [25] have been employed. In fact, these approaches were primarily used in the pre-NGS era, when the focus was on genome-wide genetic erosion, a target that is congruent with their neutral nature and the random location within the genome.

More recently, novel markers (i.e., single nucleotide polymorphisms, SNPs) derived from the application of "omics" techniques, such as restriction site associated DNA sequencing (RAD-seq) or whole-genome sequencing, have been used in evolutionary toxicology and provide a finer resolution for the detection of contaminant-induced selection [26–28]. A specific review on the pros and cons of "omics" techniques in evolutionary toxicology has recently been published [13]. Nevertheless, the potential applications of SNPs in this field are still only partially exploited. One of the main strengths of genomic approaches in comparison with microsatellites is the possibility to better investigate adaptive genetic variation by scanning for contaminant-induced selection. The common endpoints of genetic variation employed in evolutionary toxicology are the same as those used in conservation and population genetics, including heterozygosity, allele richness, Hardy-Weinberg deviations, and fixation indices. Interesting reflections are possible concerning the transferability of these endpoints to the assessment of contaminant-induced effects. The temporal and spatial scales in ecotoxicological assessments are usually more restricted than in classical population genetics studies, where the divergences are commonly evaluated in tens or hundreds of generations or on spatial distances related to relevant natural barriers. Therefore, in many cases, evolutionary toxicological responses, when detectable, can be described only as rapid changes in relation to the expected inertia of classic population genetics variations. Even if scientists now agree that contemporary evolutionary changes occurring in a few generations are not occasional but are widespread under natural or human-induced pressures [12,29], the detection of such changes using conventional experimental designs and markers is still challenging, even in the "omics" world.

In this review, we used a simulation approach to evaluate the suitability of standard experimental designs employed in population genetics and genomics with specific reference to rapid evolutionary changes. From this perspective, the implementation of powerful statistical approaches, such as approximate Bayesian computation (ABC) [30], coupled with the increase in polymorphic markers warranted by next-generation sequencing (NGS) techniques, may prove useful either in the planning phase of the assessment or in the inference of processes [31,32]. Indeed, as an example, we verified whether ABC is able to discriminate a scenario of rapid bottleneck with respect to stationarity by creating experimental designs commonly employed in population genetics and genomic surveys. We simulated genetic variability in a large population (N_e = 3000) subject to a strong (N_e = 100) or soft (N_e = 1000) bottleneck occurring abruptly in 14 generations by using DIYABC 2.1.0 [33] (details in Supplementary Material 1). Then, we verified the confidence in the scenario choice on the basis of the variability provided by 20 or 50 microsatellite loci and 2000 SNP loci. The results, expressed as type I (the probability of incorrectly assigning stationarity in the case of a true bottleneck) and type II (the probability of incorrectly assigning a bottleneck in the case of true stationarity) errors, are reported in Table 1, and indicate unexceptional performances of ABC in both strong and soft bottlenecks with both marker typologies. Only type II errors showed a clear reduction with increasing loci and the magnitude of the bottleneck, and achieved acceptable values. We also simulated the less common case of repeated temporal samples across the 14 generations (50 loci in seven replicate samples) of bottleneck and, under these conditions, both type I and type II error rates were lower, and ABC appeared to be a potentially suitable tool for optimizing the design of evolutionary toxicology studies (Table 1).

Table 1. Type I and type II error rates of scenario choices performed by approximate Bayesian computation analysis, as obtained by comparing two simulated scenarios of variable bottleneck intensity with respect to a stationary population. Different combinations of marker typology, number of loci and the availability of temporally repeated samples were evaluated.

Markers			Errors			
			Low Bottleneck		High Bottleneck	
Microsatellites	20 loci		Type I	0.36	Type I	0.43
			Type II	0.36	Type II	0.34
	50 loci		Type I	0.43	Type I	0.35
			Type II	0.31	Type II	0.09
	50 loci (rep. samples)		Type I	0.20	Type I	0.18
			Type II	0.13	Type II	0.12
Single nucleotide polymorphisms (SNPs)	2000 loci		Type I	0.38	Type I	0.46
			Type II	0.17	Type II	0.14

We also performed a non-Bayesian simulation analysis by using the online optimization tool SPOTG developed by Hoban and colleagues [34] by choosing the temporal module, which allows testing the reliability and resolution of the statistics when past (pre-bottleneck) and present (post-bottleneck) samples are available. We tested the response of statistics, such as the number of polymorphic loci, number of alleles, and heterozygosity in the same bottleneck scenarios, described above by varying the number of microsatellite loci (10, 20, and 50) and the number of specimens for each of the two samples (20, 50, and 100). The final analysis indicated a "power" of detection for each tested experimental design expressed as the percentage of simulated bottleneck datasets that fall within the lowest 5% of the distribution of a simulated constant population. The simulation results indicated a positive relationship between the magnitude of the bottleneck and the power of detection (Supplementary Materials Tables S1–S3). Regardless of the experimental design employed, the soft bottleneck could barely be discriminated, with a maximum power of 69% when considering the combination of the maximum number of loci and samples. More specifically, the number of markers were inversely correlated with the variance of the statistics; hence, at least 50 microsatellite loci were necessary to achieve low variance and high power (<90% for the strong bottleneck scenario). The number of samples was a less influential variable provided that at least 50 specimens per sample (particularly for contemporary samples) were genotyped. Among the statistics, the number of alleles was generally the most informative variable, but whenever the sample dimension between temporal samples was unequal, normalization by means of a rarefaction analysis was strongly recommended [35].

Here, we have provided only rough guidelines that can be finely tuned for each specific case and used to optimize the experimental plan. For example, the initial N_e may have a significant influence on the capacity for discriminating a bottleneck (i.e., genetic erosion) because large, stable populations with larger N_e are better able to buffer genetic bottlenecks than small, inbred ones. Nevertheless, two main conclusions can be drawn from this set of simulations. First, the experimental design required to properly analyze a typical, rapid change in genetic variability due to contaminant effects on fitness is demanding, even given the availability of hundreds or thousands of markers available through NGS. In many cases where changes in genetic variability due to contaminant exposure could not be detected, the lack of power or resolution of the chosen experimental design is a likely cause. Second, the availability of temporally repeated samples significantly enhances the performance of the experimental design.

When the research focus is more stringently narrowed towards the detection of adaptation to contaminants, specific molecular analyses such as whole-genome sequencing or RAD-seq become the current gold standard [13,36] and overcome most of the limitations of neutral or nearly neutral markers. In this regard, it is worth mentioning that loci under selection by contaminants usually represent a small proportion (1–6%) of the scanned genome [37–39], a fact that encourages the use of a wide genomic scan to identify true selected loci, particularly in natural multi-stressed systems. Nevertheless, scanning for the presence of genome regions involved in adaptation is still a challenging task [40–44]. There are still technical limitations on the capacity of genomic approaches to cover all of the regions involved in adaptation [44]. For example, it has been estimated that a RAD-seq coverage with SNPs occurring every ~5000 bp would provide a maximum coverage of ~5% of the genome [41]. Moreover, approaches that imply the enrichment of preferentially coding regions potentially increase the probability of detecting a selected region, and usually discard repetitive regions and enhancers from analysis or do not properly consider duplication events. These elements are typically the basis of gene regulation, a functional property on which adaptation to pollution may often depend [14,43,45].

Other limitations may arise from the incorrect choice of a null population model, on which the detection of selected loci is based in many available approaches and software programs. Demographic properties and events, such as highly structured populations, spatial autocorrelation of allele frequencies, or rapid range expansion, may determine deviations from the models employed by some of the most commonly used software, amplifying their error rates [41]. Moreover, background selection is known to mimic and confound the identification of positive selection [40]. Thus, the possibility of background selection should be taken into account [46], together with the structure and demographic history of a studied population, in the definition of ad hoc null models [40]. Different approaches that attempt to mitigate or overcome these issues have been proposed (see an exhaustive review in Reference [41]). Among those approaches commonly adopted in evolutionary toxicology, genetic differentiation outlier tests aim to distinguish outlier divergent loci from the average divergence of all loci, most of which are assumed to be neutral [28,47], under the assumption that different environmental conditions (i.e., stressed vs. non-stressed) enhance divergence in loci that determine sensitive phenotypes.

Genetic-environment association tests seek atypically high correlations between allele frequencies and the occurrence of contaminants (or other stressors) in exposed and non-exposed populations [36].

Specific software packages that allow, as much as possible, the shaping of null models to mirror real case studies are now available, either by adopting genetic differentiation outlier tests (e.g., [48,49]) or appropriate null models (e.g., [50,51]).

Approaches are also available that allow the simulation of datasets, taking into account different demographic events, structures, recombination maps, and different models of selection [52,53]. As in the case of a bottleneck, genotype datasets based on different experimental designs can be simulated and the conditions favoring the detectability of rapid adaptation can be assessed. To date, specific analyses from this perspective have not been published. Moreover, the implementation of selective effects on genetic variability in the framework of tools such as ABC is possible [54,55], even if it is not yet easy to standardize.

3. Paradigmatic Case Studies

3.1. Application in the Field

Early in the history of evolutionary toxicology, an extensive review of a large set of studies in aquatic environments was published in 2001 [16], emphasizing the promising perspectives for evolutionary ecotoxicology as well as warning about the need for a careful experimental setting to disentangle real patterns among the many co-occurring factors that shape genetic variability. Eighteen years later, with the increasing productivity of NGS techniques and the improved computational power of modern processors, many of the limits highlighted in that review have been scaled down. Nevertheless, many suggestions and precautions remain valid. First, although the implementation of NGS techniques has tremendously increased the information available from genetic datasets, the influence of many other factors that naturally act on genetic variability and structure, such as mutation, recombination, selection, and the genetic architecture of adaptive traits, is still not easily managed, either because of insufficient coverage of the genome or the lack of suitable data analysis techniques that are able to manage such a complexity of variables [56]. Consequently, in some cases a narrower focus on an a priori selected set of candidate genes can be justified to limit the effects of confounding factors.

Another basic approach of field studies is to complement genetic data with a careful and comprehensive assessment of exposure to ensure a representative number of reference and contaminated sites, which allows improved statistical representativeness, and to measure other life trait endpoints, biomarkers, or DNA damage descriptors. Moreover, a precise spatial characterization of stressors (including contaminant distribution and concentrations) is as fundamental as a wide coverage of the genome because the association between exposure to stressors and genome properties is the basis of most of the inferences in evolutionary toxicology [41].

Indeed, differences in genetic variability between impacted and non-impacted populations may not be solely due to contamination. Such changes may be induced by other, often predominant, factors, such as gene flow, random genetic drift, inbreeding, population density, habitat complexity, natural selection, and mutation [57–59]. The condition, not easily fulfilled, to infer a concordant pattern of changes in genetic variability (e.g., selection on specific pathways) in replicated cases of populations exposed to the same contaminants would certainly strengthen the identification of a causative role of contaminants in this regard. A complete list of all published papers on this topic is outside the scope of the present review; therefore, we focus on a few case studies that we consider paradigmatic in this field of study and that have been extensively investigated from multiple perspectives. The choice of these cases allows us to provide a survey of the potential for evolutionary toxicology to implement a risk assessment procedure (Table 2). Many other similar and equally sound studies have been conducted in the field of aquatic evolutionary toxicology.

Table 2. Field and laboratory case studies described in the review.

	Organism	Contaminant	Genetic Assay	Strengths	Opportunities	Reference
Field case studies	*Perca flavescens*	Metals (Cu, Cd)	Microsatellites, de novo transcriptome scan, microarrays	Genetic erosion and selection at unconventional targets were both detected	Test for adaptation costs toward other stressors could be performed	[10,28,60]
	Daphnia longispina	Acidity, metals	AFLP microsatellites, Comet test	Tight implementation of quantitative, population genetics, and phenotype responses	Only traditional genetics approaches were employed, no Next Generation Sequencing (NGS) data	[11]
	Solea solea	Complex mixture of pesticides and organic pollutants derived from agricultural land drainage	Microsatellites	One of the few studies finding a heterozygosity-fitness correlation, although at a single locus	Low number of samples and markers, inferences at polygenic level would have likely benefitted from NGS	[61]
	Fundulus heteroclitus	Dioxin-like contaminants	Transcriptomics, genomics, quantitative genetics	Many different approaches converge in demonstrating adaptation to contaminants and pathways affected	Long-term effects (i.e., genetic erosion, recovery from adaptation) are still uncertain and not predictable	[62]
	Gammarus pulex	Effluent of wastewater treatment plants and man-made barriers	Microsatellites	Synergic effects of habitat fragmentation, mutagenic compounds, and contaminants exposure were identified	The role of de novo mutations vs. standing genetic variation in adaptation to contaminants (e.g., with genomic approaches) could be evaluated	[63]
	Anguilla anguilla and *Anguilla rostrata*	PCBs, DDTs, metals	SNPs generated by restriction site associated DNA sequencing (RAD-seq)	Polygenic adaptation was demonstrated using RAD-seq polymorphisms analyzed by the Random Forest technique	The hypothesis of long-term genetic erosion due to contaminants could be tested	[36]
Laboratory case studies	*Heterandria formosa*	Cadmium	Microsatellites	A link between adaptation and overall genetic erosion was found; limits of closed multigeneration tests were undiscovered	Genes underpinning rapid adaptation cannot be investigated by using neutral markers	[64]
	Chironomus riparius	Tributyltin (TBT)	Microsatellites (multigenerational approach)	Effects on genetic variability were diversified at different concentrations of TBT; repeated temporal samples	Low number of neutral markers, the complex of genes adapted in relation to TBT concentration could be unveiled by integrating NGS	[65,66]
	Chironomus riparius	Perfluorinated compounds	Microsatellites (multigenerational approach)	Integration of coalescent simulations approach in evolutionary toxicology demonstrating mutation rate increase; repeated temporal samples	Low number of markers, selection could not be tested or identified, NGS approaches could be implemented at this scale	[31]

Different approaches were used by Bourret et al. [10] and by Belanger et al. [28] to determine the effects of copper and cadmium on yellow perch (*Perca flavescens*) populations. In the first study, the authors used microsatellites to investigate whether there were significant differences in levels of genetic diversity in yellow perch populations inhabiting lakes with contrasting metal levels and whether individual genetic diversity was associated with the concentration of accumulated metals within individual fish. Microsatellites were applied to individuals collected from lakes in Canadian mining regions representing a gradient of metal (copper and cadmium) contamination. The patterns of contamination within lakes and the effects on major life traits and physiological responses in yellow perch have been previously characterized in detail [67–69], with the finding that regional and seasonal factors, together with contamination, acted to shape responses. The authors not only found genetic differences between individuals from different lakes, but also found that the overall genetic diversity decreased along a gradient of increasing cadmium contamination and that copper contamination may be involved in reducing genetic diversity. Moreover, they also found that although long-term exposure to metals resulted in genetic erosion at the population level, individual genetic diversity was higher in more tolerant individuals, suggesting that less inbred individuals may be favored by selection regardless of the trend in genetic variation within a population. Five years later, Belanger-Deschenes and co-authors used an SNP genome scan to demonstrate the adaptation to Cd at two candidate loci in the same exposed populations of *P. flavescens* [28], which was reflected in a faster lifecycle leading to early sexual maturation as well as the mitigation of Cd inhibition for at least one Cd-dependent pathway. Together, these results indicated that a selective response to contamination had been sufficiently influential to reduce genetic diversity and that the long-term capacity of populations to respond to future environmental change was likely compromised.

More recently, a transcriptomics analysis of the same populations of yellow perch revealed different responses when an exposed population was transferred to a clean environment, or vice versa [60]. The analysis highlighted that specific biochemical pathways in fish were activated by contaminants when passing from clean to polluted lakes, including the under-transcription of genes involved in aerobic metabolism, the over-transcription of genes involved in protein folding, and the inhibition of the immune system. In contrast, the absence of transcriptional responses in fish moving from a polluted lake to a clean environment suggests that adaptation rather than phenotypic plasticity is the basis of their increased tolerance. Finally, the long-term potential of adapted populations to cope with other stressors, particularly infectious diseases, can be compromised.

Another noteworthy case study involved killifish (*Fundulus heteroclitus*) populations that had adapted to heavy pollution in some northern American estuaries [70]. A detailed review of this case has already been published (Reference [62] and the literature cited therein). Briefly, this study represents a fruitful integration of different approaches. Some approaches, such as quantitative genetics, physiology, and development biology, primarily contributed to the identification of adaptation and the main target of selection, which was located in the aryl hydrocarbon receptor (AHR) pathway. Other approaches, such as transcriptomics and genomics, confirmed the inferences and detected a wide range of other pathways implied in adaptation, often population-specific, which may represent responses to local stressors as well as compensatory effects acting differently in each population. Overall, this case study demonstrated that in the presence of high genetic diversity (i.e., large size and high stability of populations) and strong selective pressures, convergent adaptation toward the most influential phenotypic traits is likely.

The case of continuous acid mine drainage (AMD) on a reservoir in Portugal (Chança River Reservoir, Portugal), which carries high concentrations of many trace elements, was studied to test the presence of genetic erosion and selection on *Daphnia longispina* by comparing local reference and impacted populations (see Ribeiro et al. for a comprehensive review [11]). The integration of quantitative genetics, an assessment of many fitness endpoints and population genetics based on AFLP and microsatellites, allowed fine discrimination of the main genetic impacts and the pattern of responses of the exposed population. Specifically, genetic erosion due to the elimination of the two genotypes most sensitive to a 3% AMD pulse, and due to the avoidance of contaminated areas by the most sensitive genotypes was demonstrated in laboratory testing. A quantitative genetics approach was also employed to demonstrate the genetic basis of tolerance to lethal levels of AMD in genotypes sampled from the contaminated areas, indicating the presence of microevolution. A confirmation of this pattern of genetic erosion (i.e., the disappearance of less resistant lineages in exposed populations) was further demonstrated in the field, allowing the authors to exclude the alternative hypothesis of a spread of newly originated genotypes that conferred resistance. Interestingly, resistance was demonstrated only for the lethal concentration of AMD, but not for sublethal concentrations. Indeed, an analysis of reproductive or feeding life-traits failed to highlight significantly higher fitness in contaminated populations after exposure to sublethal concentrations of AMD or copper. A characterization of neutral genetic diversity in exposed and reference populations failed to prove genetic erosion, even when a significant divergence was detected between them. Gene flow from un-impacted areas of the reservoir and increased mutation or sexual reproduction were among the factors considered to be responsible for this pattern. A later study found mutagenic effects (micronuclei and chromosomal breaks) at 0.1% AMD [71].

Another study performed in a contaminated estuarine environment was conducted by Guinand et al., who applied a microsatellite approach to young-of-the-year sole (*Solea solea*) [61]. Sole was selected as the test organism because flatfish exploit estuaries as nursery habitats over long periods during their juvenile phase, and juveniles fail to escape pollution conditions because they are sedentary. Populations sampled at reference sites were compared with three populations sampled in estuaries contaminated by complex mixtures of pesticides and organic pollutants derived from agricultural land drainages. Using 15 genetic markers, the authors demonstrated that in heavily polluted estuaries, most of the variables were significantly different. Genetic variation was shown to be linked to selective and adaptive variation, implying a limited number of loci. Locus-specific heterozygosity-fitness correlations were found for two loci, one of which is implied in heavy metal detoxification, which was indicative of ongoing adaptive selection in sole in response to contamination. These results highlight one of the conceptual pillars of evolutionary toxicology, namely that human activities impact not only the ecological responses of species but also their evolutionary potential to cope with contamination.

Microsatellites were also used also by Inostroza et al. [63] to test the loss of biodiversity in an aquatic ecosystem impacted by mutagens from wastewaters, weirs, and other stressors. In this study, the test organism was *Gammarus pulex*, sampled along a single river that possessed several chemical sources and man-made barriers. Exposure to chemical pollution alone and in combination with the presence of weirs resulted in a depression of allelic richness in native populations. The input of mutagenic compounds resulted in a strong increase in private alleles across the affected populations, and the presence of weirs along the river disrupted migration along the river and thus the gene flow between up- and downstream individuals. This study demonstrated two of the four cornerstones of evolutionary toxicology [7] in a single case: genetic erosion due to the synergistic effect of habitat fragmentation and contaminant exposure, and an increased mutation rate mirrored in the high frequency of private alleles. The trade-off between these two counteracting drivers is intriguing, and deserves specific in-depth investigations into the genes, particularly by evaluating the relative role of standing genetic variation vs. de novo mutations in the adaptation to contaminants.

Finally, a recent study demonstrated the power of the genomic approach coupled with advanced statistical treatments to infer the effects of contaminants acting on a polygenic base [36]. This is a primary strength of this work because the detection of polygenic adaptation is a challenging task [72]. Genomic RAD-seq assays were employed to test for the presence of polygenic selection in North Atlantic eels, two species that are considered to have a high extinction risk worldwide [73]. The two species, *Anguilla anguilla* (the European eel) and *Anguilla rostrata* (the American eel) were used as test replicates and a Random Forest search identified a total of 142 and 141 co-varying loci, respectively, that discriminated "polluted" from "control" populations. Overall, subtle allelic frequency changes were found to be associated with the bioaccumulation of PCB153, pp'-DDE, and selenium.

Moreover, the authors succeeded in annotating the discriminant markers and concluded that the regulation, absorption, and transport of sterols appeared to play a major role in the differential survival of eels in polluted environments.

3.2. Laboratory Applications

Regarding the application of genetic assays in laboratory tests, few works have evaluated long multigenerational exposures. Indeed, in the environment, organisms are often exposed to substances over a time scale that is typically impossible to mimic in the laboratory. To capture population genetic effects, the exposure time must be extended over multiple generations, which is a significant challenge in laboratory ecotoxicological studies. Multigenerational tests can be successfully conducted in only a few species that are easy to breed. However, a well-planned multigenerational test may significantly accelerate processes, such as selection, adaptation, or inbreeding, accomplishing in the laboratory what requires a relatively long time in the field. A multigenerational test, as explained below, can be significantly more realistic from an ecological perspective than normal single generation tests. Indeed, as in the traditional ecotoxicological approach, the controlled conditions of the laboratory test could favor a univocal data interpretation, whereas field studies can be affected by all of the (uncontrolled) variables that are present in a natural ecosystem. Finally, the ability to analyze repeated temporal samples collected during the experiment can significantly improve the resolution of the experimental design, as described in Section 3.1.

Most of the multigenerational experiments published to date are relatively recent and all employ the microsatellite technique. Athrey et al. [64] applied the technique to the least killifish *Heterandria formosa* to assess the loss of genetic variation in laboratory populations after eight generations of strong selection for an increased resistance to cadmium. The authors compared genetic variation between three selected and three control laboratory populations and between these laboratory populations and the source population by maintaining the same number of breeders (60 specimens) and sex ratio (1:1) between replicates and across generations. A previous analysis found a rapid six-fold increase in resistance and specific fitness costs associated with this adaptation to lethal Cd concentrations. Heterozygosity was lower in each selected population than in its paired control population, and this difference was statistically significant in two of the three comparisons. Using this genetic approach, the authors demonstrated that adaptation to environmental contaminants can result in an overall loss of genetic variation, and they related this loss to an increased variance in reproduction in exposed populations due to a probable differential survival of families in response to Cd exposure. This study emphasized that adaptation to lethal levels of a contaminant can induce overall genetic erosion in the absence of external factors counterbalancing the loss of genetic variability. Nevertheless, specific investigations focusing on the genes involved in the adaptation to Cd were not possible based only on seven microsatellite loci. This study also addressed one of the critical points of multigenerational studies: the unavoidable loss of genetic variation in all test populations due to the genetic effective population size imposed by the experimental setting and the absence of external gene flow. In this specific case, the genetic effective population size was estimated to be ~20% of the census population size in all treatments after eight generations, and the loss of genetic variability due to rearing conditions was significant. Indeed, in all experimental lines, only about two alleles/loci were left and heterozygosity was strongly reduced (32.11% and 37.8% for controls and exposed lines, respectively).

The loss of genetic variation resulting from breeding populations in the laboratory demonstrates that it is important to maintain a large population size and that the potential loss of genetic variation in laboratory populations should be taken into consideration when extrapolating from laboratory to natural populations. Again, the use of simulation approaches could assist in determining the experimental conditions suitable to maintain sufficient genetic variability despite genetic drift, generating a null scenario of random mating, and thus enabling exposed and unexposed populations to be compared.

Whereas adaptation to lethal levels of contaminants was the focus of the work by Athrey et al., other authors have adopted the multigenerational (12 generations) approach to simulate a long-term chronic and sublethal exposure and to investigate gradual population phenotypic responses to the stressor. In this regard, an interesting case study was proposed by Vogt and Novak [65,66]. In their works, the midge (*Chironomus riparius*) was used as a freshwater model organism to investigate the effects of the highly toxic biocide tributyltin (TBT) at two different sublethal, environmentally relevant concentrations. The study aim was to monitor changes in the population genetic structure as a response to a toxic stress due to adaptation processes or the reduction of neutral genetic variation. In addition to the investigation of genetic variation in the stressed population, fitness-relevant parameters (e.g., mortality and reproduction) were also monitored across all generations. Regarding genetic diversity, it is interesting that the authors found different patterns of response in populations depending on the contaminant concentrations. In the case of milder exposure (4.46 µg Sn kg^{-1} sediment, dw), no significant differences were found in the level of heterozygosity or allele richness at five microsatellite loci between the exposed and the control populations, but deviations from the Hardy-Weinberg equilibrium accumulated at two loci over the generations in the TBT-exposed populations, suggesting possible selective effects. Moreover, considerable evidence for ongoing adaptation processes was suggested by a significant tolerance to TBT in an acute test and a simultaneous increase in reproductive output in later generations.

This study was preliminary to a second study, which investigated the effects of a nearly doubled concentration of TBT [65]. The experimental plan was the same: a 12-generation test was performed on the same model organism (*C. riparius*) with the aim of revealing and measure genetic variation in relation to life trait responses. In this second study, a significantly higher decrease in overall genetic variation, in terms of heterozygosity and allele richness, was detected in TBT-exposed populations. No evidence for selection processes was detected, as no significant time trend in any life-history trait was observed, and tolerance towards TBT did not significantly change over time. However, reproductive impairments (increased mortalities and a reduced number of fertile clutches) in most generations were associated with TBT exposure.

Overall, both of the studies appeared to indicate that different intensities of the same stressor modulate the reciprocal balance between selection and genetic drift. Unfortunately, the experimental design did not allow a detailed evaluation of whether the pool of genes under selection varied significantly in the two cases or under the same experimental conditions regardless of the presence of the contaminant. Indeed, it must be remembered that microsatellites are typically neutral markers except when linked to selected traits, which is a highly improbable event given that toxicant-induced selection usually occurs in a very low percentage of loci within the genome.

The results of these studies were fundamental for ratifying the differences between single-generation and multigeneration tests. Indeed, two main lines of evidence emerged against the significance of short-term assays. First, there was high, apparently stochastic, variability in responses linked to life traits, often indicating opposite and contrasting inferences about the toxicity of contaminants from one generation to another [65]. Second, long-term significant effects on genetic variability were present despite the absence of clear phenotypic responses, which appeared to indicate that the traditional tools employed for ecological risk assessment may in many cases underestimate the long-term risk of extinction for resident populations.

In another case study, a different response at the genetic level was found by Stefani and co-workers [31], who aimed to identify the long-term effects of perfluorinated compounds (PFAS), which are very persistent contaminants of emerging concern. According to the experimental framework of Nowak et al. [58], the authors exposed some populations of *C. riparius* to 10 µg L^{-1} PFOA, PFOS, or PFBS (two replicates per treatment) for 10 generations, together with two contaminant-free controls.

For the tested substances other than PFOS, low acute toxicity was found for test organisms with standard end-points, indicating toxic concentrations significantly above the levels found in the environment. Genetic analysis demonstrated the maintenance of heterozygosity and allelic diversity in

PFOS- and PFBS-treated populations with respect to the controls across generations, whereas no effects for PFOA-treated samples were detected despite a general loss of genetic diversity in all treatments due to the experimental design. The results obtained were not indicative of genetic erosion caused by PFASs. On the contrary, a pattern of increased mutation rates emerged as the main transgenerational effect, as demonstrated by the integration of the coalescent simulation approach and the availability of temporally repeated samples and the absence of gene flow. A possible explanation is that an increased mutation rate may have been caused directly by exposure to the contaminant [6] because of the chemical properties of the sulphonate groups carried by PFOS and PFBS, or indirectly by physiological stress [74]. Indeed, exposure to PFOS and PFBS may have induced worse physiological conditions than those observed in the absence of toxicants, and thus generated an increase in mutation rate. A genetic characterization of wild populations of a caddisfly inhabiting a site highly impacted by PFOA demonstrated significant effects (i.e., an altered genetic structure) as well [75].

4. Potential Role of Evolutionary Toxicology in Ecological Risk Assessments

The introduction of an evolutionary perspective to the ecological risk assessment (ERA) of chemical contaminants has received limited attention to date, although the integration of evolutionary toxicology with regulatory procedures has been recommended since the dawn of the discipline [5,7,16]. Indeed, although a long-term risk for populations linked to the alteration of genetic properties was considered plausible, disentangling this detrimental process from other co-occurring stressors has nevertheless been challenging in many cases [76]. This was in part due to the awareness that the available genetic approaches were scarcely standardizable, often requiring relevant analytical and time resources, and their output was challenging to convert to a risk scale. Here, we illustrate the potential for ABC simulations to optimize the design of population genetic studies to detect chemical impacts, including both bottlenecks and selection [30,34,52,53,77] (Figure 1). This would allow testing the feasibility of population genetics or genomic surveys in specific contexts to evaluate and quantify the costs necessary to reach significant and acceptable inferences and, at the end of the assay, to provide responses by testing alternative scenarios [31] based on a set of the most informative statistics.

Recently, thanks to the improvement in genetic techniques, more and more studies have succeeded in demonstrating a fundamental role of evolutionary dynamics in determining responses to environmental stressors, including chemical pollution [78,79]. A paradigmatic case study involves the adaptation of Atlantic killifish (*Fundulus heteroclitus*) to a wide range of toxicants, particularly PCBs, in northern American estuaries [62]. In addition to demonstrating a rapid convergent adaptation that favors the inhibition of the AHR pathway, the primary target of dioxine-like compounds, this study provided the opportunity to draw useful, general guidelines that can be extended to other potential cases and introduced to an ERA framework. First, the intrinsic properties of the exposed populations, such as a high population size, short generation time, and high genetic diversity, increase the probability of a rapid adaptation to toxicants, whereas the opposite conditions could be related to a higher stochastic extinction risk [46,72,80]. One might assume (often incorrectly) that species with a higher probability of adaptation would likely be of less concern for protection, whereas rare and endangered taxa would be more prone to extinction when they confront a relevant chemical contamination. Nevertheless, it should be considered that even if adaptation apparently acts as a rescue strategy, it often carries fitness costs, particularly in terms of a decreased resistance to other stressors, which could indirectly impair the long-term persistence of even abundant and prolific species [81–83]. Moreover, abundant species often play a key role in maintaining ecosystem functionality, and their reduced fitness or demographic contraction may significantly impair the provision of ecosystem services [84]. Overall, this guideline can have resounding relevance for conservation prioritization and risk assessment optimization, aiding the choice of suitable and sensitive model taxa for evaluating either short-term risks (i.e., extinction risks via population viability analysis) in the case of endangered and small populations, which likely have limited potential to adapt rapidly, or long-term risks (i.e., fitness costs to adaptation) in the case of abundant and prolific populations.

Moreover, adaptations to a single or a few prevalent compounds were more likely, more rapid, and less demanding than to a wider mixture of less concentrated pollutants in terms of energetic trade-offs. For example, exposure to pulsed, highly concentrated, specific target toxicants (e.g., pesticides) likely drives the selection of simple phenotypes related to a single or a few physiological targets, whereas chronic exposure to sublethal mixtures of contaminants with multiple modes of action requires the selection of complex, multidimensional phenotypes, which is a less likely process that requires a long time to evolve [62]. In the latter case, further investigations of evolutionary implications may be focused primarily on the first category of contaminants in the context of the process of prioritization of substances for surveillance monitoring.

A corollary of this principle is that the severity of the exposure influences the complexity of genes involved in adaptation. Whenever the intensity of an environmental stressor exceeds the upper tail of the sensitivity distribution of phenotypes within a population, few genes with large effects are typically involved in adaptation. In the case of a milder, prolonged, and gradual intensity of perturbation, polygenic adaptation becomes more likely. Again, it is worth noting that adaptation to a simple complex of genes does not imply that long-term risks may be less relevant than those under polygenic adaptation. Indeed, pleiotropy is a common property of genomes [85], and the indirect effects of selection on numerous phenotypic traits, even those related to a single gene, are frequent, often leading to compensatory adaptations in other genes (assuming that physiological adjustments have failed to restore homeostasis) [86,87].

It is now evident that the long-term adaptation to toxicants is a common phenomenon in nature, and effective mechanisms of resistance and detoxification are often shared between related phylogenetic lineages [88]. This awareness has raised the possibility that the sensitivity of taxa to contaminants could be in some way be predicted by their phylogenetic affinity. For example, some authors [85] provided evidence of a significant correlation between the sensitivity of taxa to chloride, a typical natural stressor that organisms have faced since the transition from marine to freshwater habitats, and their phylogenetic relatedness. Hence, the possibility of estimating toxicity thresholds for untested taxa by extrapolating from the available estimates of related species appears a reasonable approach in ERA to increase ecological realism and to escape from the intrinsic limits of the model species approach [89].

Even the short-term adaptation to toxicants by resident species may have a relevant influence on the ERA. It is well known that the current normative approach has limits in assessing long-term effects, which can persist after the exposure has ceased, particularly under multiple stressor conditions [8,76]. Indeed, adaptation to toxicants by wild populations may provide sensitivity to the selective factor or to other stressors, which can be significantly different from that estimated in laboratory assays, which are the basis for the derivation of toxicological thresholds. For example, variation in the response to chloride exposure by adapted and non-adapted conspecific wild populations of two amphibians was large, and exceeded the range of variability estimated among other amphibian species [85]. Consequently, a reasonable improvement in the ERA should not overlook the probability of adaptation by resident populations to pre-existing stressors (Figure 1), and precautionary approaches should be adopted whenever specific assays are not available or conducted.

5. Conclusions

Evidence that chemical contaminants may have direct or indirect impacts on the genetic variability of exposed populations is now growing. Moreover, thanks to the integration of NGS approaches, mechanisms and processes of adaptation have been revealed. The negative impacts on the long-term probability of survival of adapted populations are no longer described only as theoretical possibilities, but rather can be tested and hopefully predicted (see examples in References [62,90]).

As demonstrated in the studies reviewed above, the use of a genetic approach can highlight the differences between exposed and reference populations both in laboratory tests and in field research, raising concerns about contaminant effects and the fate of the populations involved. Nevertheless, there is also some evidence of a relevant counteracting effect of natural dynamics, such as migration and gene flow, or different synergistic, neutral, or antagonistic interactions of adaptation with other stressors. This is probably one of the most important aspects that should be addressed in the immediate future, with the ultimate aim of improving the predictability of the long-term fate of exposed populations. This consideration should not, however, be assumed to be indicative of a limited relevance of evolutionary changes in the context of ecotoxicological risk assessment. Indeed, the reviewed case studies demonstrated that the investigation of adaptation to contamination, with its intrinsic drawbacks, is greatly improved by the implementation of genomic scan techniques, which are just starting to be massively employed in this discipline.

As suggested by Bourret et al., our review highlights several examples indicating that evolutionary change may occur more rapidly in our lifetime [10] and provides more evidence that human activities are not only affecting the demography and the ecology of wild species, but also their evolutionary trajectory [76,79]. This statement alone should be considered sufficient to integrate evolutionary approaches into environmental assessment procedures.

Supplementary Materials: The following are available online at http://www.mdpi.com/2073-4441/10/4/490/s1.

Acknowledgments: The authors thank the four anonymous reviewers for their fruitful advice on a previous version of the manuscript.

Author Contributions: Marianna Rusconi and Fabrizio Stefani conceived and wrote the review; Roberta Bettinetti and Stefano Polesello gave a critical contribution to the discussion about the possible implementation of evolutionary toxicology in ecotoxicological risk assessment.

Conflicts of Interest: The authors declare no conflict of interest.

References

1. Simmons, D.B.D.; Benskin, J.P.; Cosgrove, J.R.; Duncker, B.P.; Ekman, D.R.; Martyniuk, C.J.; Sherry, J.P. Omics for aquatic ecotoxicology: Control of extraneous variability to enhance the analysis of environmental effects. *Environ. Toxicol. Chem.* **2015**, 34, 1693–1704. [CrossRef] [PubMed]

2. Klerks, P.L.; Xie, L.; Levinton, J.S. Quantitative genetics approaches to study evolutionary processes in ecotoxicology; a perspective from research on the evolution of resistance. *Ecotoxicology* **2011**, 20, 513–523. [CrossRef] [PubMed]

3. Brown, A.R.; Hosken, D.J.; Balloux, F.; Bickley, L.K.; LePage, G.; Owen, S.F.; Hetheridge, M.J.; Tyler, C.R. Genetic variation, inbreeding and chemical exposure—Combined effects in wildlife and critical considerations for ecotoxicology. *Philos. Trans. R. Soc. B Biol. Sci.* **2009**, 364, 3377–3390. [CrossRef] [PubMed]

4. Anderson, C.; Cunha, L.; Sechi, P.; Kille, P.; Spurgeon, D. Genetic variation in populations of the earthworm, *Lumbricus rubellus*, across contaminated mine sites. *BMC Genet.* **2017**, 18, 1–13. [CrossRef] [PubMed]

5. Coutellec, M.A.; Barata, C. An introduction to evolutionary processes in ecotoxicology. *Ecotoxicology* **2011**, 20, 493–496. [CrossRef] [PubMed]

6. Bickham, J.W.; Smolen, M.J. Somatic and heritable effects of environmental genotoxins and the emergence of evolutionary toxicology. *Environ. Health Perspect.* **1994**, 102, 25–28. [CrossRef] [PubMed]

7. Bickham, J.W. The four cornerstones of evolutionary toxicology. *Ecotoxicology* **2011**, 20, 497–502. [CrossRef] [PubMed]

8. Stoks, R.; Debecker, S.; Van, K.D.; Janssens, L. Integrating ecology and evolution in aquatic toxicology: Insights from damselflies. *Freshw. Sci.* **2015**, 34, 1032–1039. [CrossRef]

9. Bickham, J. Effects of chemical contaminants on genetic diversity in natural populations: Implications for biomonitoring and ecotoxicology. *Mutat. Res.* **2000**, 463, 33–51. [CrossRef]

10. Bourret, V.; Couture, P.; Campbell, P.G.C.; Bernatchez, L. Evolutionary ecotoxicology of wild yellow perch (*Perca flavescens*) populations chronically exposed to a polymetallic gradient. *Aquat. Toxicol.* **2008**, 86, 76–90. [CrossRef] [PubMed]

11. Ribeiro, R.; Baird, D.J.; Soares, A.M.V.M.; Lopes, I. Contaminant driven genetic erosion: A case study with *Daphnia longispina*. *Environ. Toxicol. Chem.* **2012**, 31, 977–982. [CrossRef] [PubMed]

12. Brady, S.P.; Monosson, E.; Matson, C.W.; Bickham, J.W. Evolutionary toxicology: Toward a unified understanding of life's response to toxic chemicals. *Evolut. Appl.* **2017**, *10*, 745–751. [CrossRef] [PubMed]

13. Oziolor, E.M.; Bickham, J.W.; Matson, C.W. Evolutionary toxicology in an omics world. *Evolut. Appl.* **2017**, *10*, 752–761. [CrossRef] [PubMed]

14. Shaw, J.R.; Hampton, T.H.; King, B.L.; Whitehead, A.; Galvez, F.; Gross, R.H.; Keith, N.; Notch, E.; Jung, D.; Glaholt, S.P.; et al. Natural selection canalizes expression variation of environmentally induced plasticity-enabling genes. *Mol. Biol. Evolut.* **2014**, *31*, 3002–3015. [CrossRef] [PubMed]

15. Wernersson, A.S.; Carere, M.; Maggi, C.; Tusil, P.; Soldan, P.; James, A.; Sanchez, W.; Dulio, V.; Broeg, K.; Reifferscheid, G.; et al. The European technical report on aquatic effect-based monitoring tools under the water framework directive. *Environ. Sci. Eur.* **2015**, *27*, 7. [CrossRef]

16. Belfiore, N.M.; Anderson, S.L. Effects of contaminants on genetic patterns in aquatic organisms: A review. *Mutat. Res. Rev. Mutat. Res.* **2001**, *489*, 97–122. [CrossRef]

17. Gienapp, P.; Fior, S.; Guillaume, F.; Lasky, J.R.; Sork, V.L.; Csilléry, K. Genomic quantitative genetics to study evolution in the wild. *Trends Ecol. Evolut.* **2017**, *32*, 897–908. [CrossRef] [PubMed]

18. Suarez-Ulloa, V.; Gonzalez-Romero, R.; Eirin-Lopez, J.M. Environmental epigenetics: A promising venue for developing next-generation pollution biomonitoring tools in marine invertebrates. *Mar. Pollut. Bull.* **2015**, *98*, 5–13. [CrossRef] [PubMed]

19. Peterson, E.K.; Buchwalter, D.B.; Kerby, J.L.; Lefauve, M.K.; Varian-Ramos, C.W.; Swaddle, J.P. Integrative behavioral ecotoxicology: Bringing together fields to establish new insight to behavioral ecology, toxicology, and conservation. *Curr. Zool.* **2017**, *63*, 185–194. [CrossRef] [PubMed]

20. Brander, S.M.; Biales, A.D.; Connon, R.E. The role of epigenomics in aquatic toxicology. *Environ. Toxicol. Chem.* **2017**, *36*, 2565–2573. [CrossRef] [PubMed]

21. Mussali-galante, P.; Tovar-sánchez, E.; Valverde, M.; Rojas, E. *Reviews of Environmental Contamination and Toxicology*; Springer: New York, NY, USA, 2014; Volume 227.

22. Selkoe, K.A.; Toonen, R.J. Microsatellites for ecologists: A practical guide to using and evaluating microsatellite markers. *Ecol. Lett.* **2006**, *9*, 615–629. [CrossRef] [PubMed]

23. Agnèse, J.F.; Adépo-Gourène, B.; Nyingi, D. Functional microsatellite and possible selective sweep in natural populations of the black-chinned tilapia *Sarotherodon melanotheron* (Teleostei, Cichlidae). *Mar. Genom.* **2008**, *1*, 103–107. [CrossRef] [PubMed]

24. Rengmark, A.H.; Lingaas, F. Genomic structure of the Nile tilapia (*Oreochromis niloticus*) transferrin gene and a haplotype associated with saltwater tolerance. *Aquaculture* **2007**, *272*, 146–155. [CrossRef]

25. Vos, P.; Hogers, R.; Bleeker, M.; Reijans, M.; Lee, T.V.D.; Frijters, A.; Pot, J.; Peleman, J.; Kuiper, M.; Zabeau, M.; et al. AFLP: A new technique for DNA fingerprinting. *Nucleic Acids Res.* **1995**, *23*, 4407–4414. [CrossRef] [PubMed]

26. Bouétard, A.; Côte, J.; Besnard, A.-L.; Collinet, M.; Coutellec, M.-A. Environmental versus anthropogenic effects on population adaptive divergence in the freshwater snail *Lymnaea stagnalis*. *PLoS ONE* **2014**, *9*, e106670. [CrossRef] [PubMed]

27. Bouétard, A.; Noirot, C.; Besnard, A.L.; Bouchez, O.; Choisne, D.; Robe, E.; Klopp, C.; Lagadic, L.; Coutellec, M.A. Pyrosequencing-based transcriptomic resources in the pond snail *Lymnaea stagnalis*, with a focus on genes involved in molecular response to diquat-induced stress. *Ecotoxicology* **2012**, *21*, 2222–2234. [CrossRef] [PubMed]

28. Bélanger-Deschênes, S.; Couture, P.; Campbell, P.G.C.; Bernatchez, L. Evolutionary change driven by metal exposure as revealed by coding SNP genome scan in wild yellow perch (*Perca flavescens*). *Ecotoxicology* **2013**, *22*, 938–957. [CrossRef] [PubMed]

29. Hendry, A.P.; Gotanda, K.M.; Svensson, E.I. Human influences on evolution, and the ecological and societal consequences. *Philos. Trans. R. Soc. B Biol. Sci.* **2017**, *372*, 20160028. [CrossRef] [PubMed]

30. Csilléry, K.; Blum, M.G.B.; Gaggiotti, O.E.; François, O. Approximate Bayesian Computation (ABC) in practice. *Trends Ecol. Evolut.* **2010**, *25*, 410–418. [CrossRef] [PubMed]

31. Stefani, F.; Rusconi, M.; Valsecchi, S.; Marziali, L. Evolutionary ecotoxicology of perfluoralkyl substances (PFASs) inferred from multigenerational exposure: A case study with *Chironomus riparius* (Diptera, Chironomidae). *Aquat. Toxicol.* **2014**, *156*, 41–51. [CrossRef] [PubMed]

32. Momigliano, P.; Jokinen, H.; Fraimout, A.; Florin, A.-B.; Norkko, A.; Merilä, J. Extraordinarily rapid speciation in a marine fish. *Proc. Natl. Acad. Sci. USA* **2017**, *114*, 6074–6079. [CrossRef] [PubMed]

33. Cornuet, J.; Pudlo, P.; Veyssier, J.; Dehne-garcia, A.; Marin, J.; Estoup, A.; Gautier, M.; Cnrs, U.M.R. DIYABC v2. 0: A software to make approximate Bayesian computation inferences about population history using single nucleotide polymorphism, DNA sequence and microsatellite data. *Bioinformatics* **2018**, *30*, 1187–1189. [CrossRef] [PubMed]

34. Hoban, S.; Gaggiotti, O.; Bertorelle, G. Sample Planning Optimization Tool for conservation and population Genetics (SPOTG): A software for choosing the appropriate number of markers and samples. *Methods Ecol. Evolut.* **2013**, *4*, 299–303. [CrossRef]

35. Hoban, S.; Arntzen, J.A.; Bruford, M.W.; Godoy, J.A.; Rus Hoelzel, A.; Segelbacher, G.; Vilà, C.; Bertorelle, G. Comparative evaluation of potential indicators and temporal sampling protocols for monitoring genetic erosion. *Evolut. Appl.* **2014**, *7*, 984–998. [CrossRef] [PubMed]

36. Laporte, M.; Pavey, S.A.; Rougeux, C.; Pierron, F.; Lauzent, M.; Budzinski, H.; Labadie, P.; Geneste, E.; Couture, P.; Baudrimont, M.; et al. RAD sequencing reveals within-generation polygenic selection in response to anthropogenic organic and metal contamination in North Atlantic Eels. *Mol. Ecol.* **2016**, *25*, 219–237. [CrossRef] [PubMed]

37. Bach, L.; Dahllöf, I. Local contamination in relation to population genetic diversity and resilience of an arctic marine amphipod. *Aquat. Toxicol.* **2012**, *114–115*, 58–66. [CrossRef] [PubMed]

38. Boyer, B. Genome scan in the mosquito *Aedes rusticus*: Population structure and detection of positive selection after insecticide treatment. *Mol. Ecol.* **2010**, 325–337. [CrossRef]

39. Vega-Retter, C.; Vila, I.; Véliz, D. Signatures of directional and balancing selection in the silverside *Basilichthys microlepidotus* (Teleostei: Atherinopsidae) inhabiting a polluted river. *Evolut. Biol.* **2015**, *42*, 156–168. [CrossRef]

40. Bank, C.; Ewing, G.B.; Ferrer-Admettla, A.; Foll, M.; Jensen, J.D. Thinking too positive? Revisiting current methods of population genetic selection inference. *Trends Genet.* **2014**, *30*, 540–546. [CrossRef] [PubMed]

41. Hoban, S.; Kelley, J.L.; Lotterhos, K.E.; Antolin, M.F.; Bradburd, G.; Lowry, D.B.; Poss, M.L.; Reed, L.K.; Storfer, A.; Whitlock, M.C. Finding the genomic basis of local adaptation: Pitfalls, practical solutions, and future directions. *Am. Nat.* **2016**, *188*, 379–397. [CrossRef] [PubMed]

42. Hohenlohe, P.A.; Phillips, P.C.; Cresko, W.A. Using population genomics to detect selection in natural populations: Key concepts and methodological considerations. *Int. J. Plant Sci.* **2011**, *171*, 1059–1071. [CrossRef] [PubMed]

43. Tiffin, P.; Ross-Ibarra, J. Advances and limits of using population genetics to understand local adaptation. *Trends Ecol. Evolut.* **2014**, *29*, 673–680. [CrossRef] [PubMed]

44. Merilä, J.; Hendry, A.P. Climate change, adaptation, and phenotypic plasticity: The problem and the evidence. *Evolut. Appl.* **2014**, *7*, 1–14. [CrossRef] [PubMed]

45. Whitehead, A.; Galvez, F.; Zhang, S.; Williams, L.M.; Oleksiak, M.F. Functional genomics of physiological plasticity and local adaptation in killifish. *J. Hered.* **2011**, *102*, 499–511. [CrossRef] [PubMed]

46. Comeron, J.M. Background selection as null hypothesis in population genomics: Insights and challenges from *Drosophila* studies. *Philos. Trans. R. Soc. B Biol. Sci.* **2017**, *372*, 20160471. [CrossRef] [PubMed]

47. Proestou, D.A.; Flight, P.; Champlin, D.; Nacci, D. Targeted approach to identify genetic loci associated with evolved dioxin tolerance in Atlantic Killifish (*Fundulus heteroclitus*). *BMC Evolut. Biol.* **2014**, *14*. [CrossRef] [PubMed]

48. Foll, M.; Gaggiotti, O. A genome-scan method to identify selected loci appropriate for both dominant and codominant markers: A Bayesian perspective. *Genetics* **2008**, *180*, 977–993. [CrossRef] [PubMed]

49. Vitalis, R.; Gautier, M.; Dawson, K.J.; Beaumont, M.A. Detecting and measuring selection from gene frequency data. *Genetics* **2014**, *196*, 799–817. [CrossRef] [PubMed]

50. Günther, T.; Coop, G. Robust identification of local adaptation from allele frequencies. *Genetics* **2013**, *195*, 205–220. [CrossRef] [PubMed]

51. Gautier, M. Genome-wide scan for adaptive divergence and association with population-specific covariates. *Genetics* **2015**, *201*, 1555–1579. [CrossRef] [PubMed]

52. Messer, P.W. SLiM: Simulating evolution with selection and linkage. *Genetics* **2013**, *194*, 1037–1039. [CrossRef] [PubMed]

53. Hernandez, R.D. A flexible forward simulator for populations subject to selection and demography. *Bioinformatics* **2008**, *24*, 2786–2787. [CrossRef] [PubMed]

54. Jensen, J.D.; Thornton, K.R.; Andolfatto, P. An approximate bayesian estimator suggests strong, recurrent selective sweeps in drosophila. *PLoS Genet.* **2008**, *4*. [CrossRef] [PubMed]

55. Bazin, E.; Dawson, K.J.; Beaumont, M.A. Likelihood-free inference of population structure and local adaptation in a Bayesian hierarchical model. *Genetics* **2010**, *185*, 587–602. [CrossRef] [PubMed]

56. Haasl, R.J.; Payseur, B.A. Fifteen years of genomewide scans for selection: Trends, lessons and unaddressed genetic sources of complication. *Mol. Ecol.* **2016**, 5–23. [CrossRef] [PubMed]

57. Nei, M.; Maruyama, T.; Chakraborty, R. The bottleneck effect and genetic variability in populations. *Evolution* **1975**, *29*, 1–10. [CrossRef] [PubMed]

58. Hartl, D.L. *A Primer of Population Genetics*; Sinauer Associates, Inc.: Sunderland, MA, USA, 2001; p. 221.

59. De Meeûs, T.; Michalakis, Y.; Renaud, F.; Olivieri, I. Polymorphism in heterogeneous environments, evolution of habitat selection and sympatric speciation: Soft and hard selection models. *Evolut. Ecol.* **1993**, *7*, 175–198. [CrossRef]

60. Bougas, B.; Normandeau, E.; Grasset, J.; Defo, M.A.; Campbell, P.G.C.; Couture, P.; Bernatchez, L. Transcriptional response of yellow perch to changes in ambient metal concentrations-A reciprocal field transplantation experiment. *Aquat. Toxicol.* **2016**, *173*, 132–142. [CrossRef] [PubMed]

61. Guinand, B.; Fustier, M.A.; Labonne, M.; Jourdain, E.; Calvès, I.; Quiniou, L.; Cerqueira, F.; Laroche, J. Genetic structure and heterozygosity-fitness correlation in young-of-the-year sole (*Solea solea* L.) inhabiting three contaminated West-European estuaries. *J. Sea Res.* **2013**, *80*, 35–49. [CrossRef]

62. Whitehead, A.; Clark, B.W.; Reid, N.M.; Hahn, M.E.; Nacci, D. When evolution is the solution to pollution: Key principles, and lessons from rapid repeated adaptation of killifish (*Fundulus heteroclitus*) populations. *Evolut. Appl.* **2017**, *10*, 762–783. [CrossRef] [PubMed]

63. Inostroza, P.A.; Vera-Escalona, I.; Wicht, A.J.; Krauss, M.; Brack, W.; Norf, H. Anthropogenic stressors shape genetic structure: Insights from a model freshwater population along a land use gradient. *Environ. Sci. Technol.* **2016**, *50*, 11346–11356. [CrossRef] [PubMed]

64. Athrey, N.R.G.; Leberg, P.L.; Klerks, P.L. Laboratory culturing and selection for increased resistance to cadmium reduce genetic variation in the least killifish, *Heterandria formosa*. *Environ. Toxicol. Chem.* **2007**, *26*, 1916–1921. [CrossRef] [PubMed]

65. Nowak, C.; Vogt, C.; Pfenninger, M.; Schwenk, K.; Oehlmann, J.; Streit, B.; Oetken, M. Rapid genetic erosion in pollutant-exposed experimental chironomid populations. *Environ. Pollut.* **2009**, *157*, 881–886. [CrossRef] [PubMed]

66. Vogt, C.; Nowak, C.; Barateiro, J.; Oetken, M.; Schwenk, K. Multi-generation studies with *Chironomus riparius*—Effects of low tributyltin concentrations on life history parameters and genetic diversity. *Chemosphere* **2007**, *67*, 2192–2200. [CrossRef] [PubMed]

67. Couture, P.; Busby, P.; Gauthier, C.; Rajotte, J.W.; Pyle, G.G. Seasonal and regional variations of metal contamination and condition indicators in yellow perch (*Perca flavescens*) along two polymetallic gradients. I. Factors influencing tissue metal concentrations. *Hum. Ecol. Risk Assess.* **2008**, *14*, 97–125. [CrossRef]

68. Couture, P.; Rajotte, J.W.; Pyle, G.G. Seasonal and regional variations in metal contamination and condition indicators in yellow perch (*Perca flavescens*) along two polymetallic gradients. III. Energetic and physiological indicators. *Hum. Ecol. Risk Assess.* **2008**, *14*, 146–165. [CrossRef]

69. Pyle, G.; Busby, P.; Gauthier, C.; Rajotte, J.; Couture, P. Seasonal and regional variations in metal contamination and condition indicators in yellow perch (*Perca flavescens*) along two polymetallic gradients. II. Growth patterns, longevity, and condition. *Hum. Ecol. Risk Assess.* **2008**, *14*, 126–145. [CrossRef]

70. Reid, N.M.; Proestou, D.A.; Clark, B.W.; Warren, W.C.; Colbourne, J.K.; Shaw, J.R.; Karchner, S.I.; Hahn, M.E.; Nacci, D.; Oleksiak, M.F.; et al. The genomic landscape of rapid repeated evolutionary adaptation to toxic pollution in wild fish. *Science* **2016**, *354*, 1305–1308. [CrossRef] [PubMed]

71. Sobral, O.; Marin-Morales, M.A.; Ribeiro, R. Could contaminant induced mutations lead to a genetic diversity overestimation? *Ecotoxicology* **2013**, *22*, 838–846. [CrossRef] [PubMed]

72. Josephs, E.B.; Stinchcombe, J.R.; Wright, S.I. What can genome-wide association studies tell us about the evolutionary forces maintaining genetic variation for quantitative traits? *New Phytol.* **2017**, *214*, 21–33. [CrossRef] [PubMed]

73. *The IUCN Red list of Threatened Species. Version 2016-3*; Version 20; IUCN (International Union for Conservation of Nature): Gland, Switzerland, 2016.

74. Baer, C.F. Does mutation rate depend on itself? *PLoS Biol.* **2008**, *6*, e52. [CrossRef] [PubMed]

75. Rusconi, M.; Marziali, L.; Stefani, F.; Valsecchi, S.; Bettinetti, R.; Mazzoni, M.; Rosignoli, F.; Polesello, S. Evaluating the impact of a fluoropolymer plant on a river macrobenthic community by a combined chemical, ecological and genetic approach. *Sci. Total Environ.* **2015**, *538*, 654–663. [CrossRef] [PubMed]

76. Ashley, M.V.; Willson, M.F.; Pergams, O.R.W.; O'Dowd, D.J.; Gende, S.M.; Brown, J.S. Evolutionarily enlightened management. *Biol. Conserv.* **2003**, *111*, 115–123. [CrossRef]

77. Laval, G.; Excoffier, L. SIMCOAL 2.0: A program to simulate genomic diversity over large recombining regions in a subdivided population with a complex history. *Bioinformatics* **2004**, *20*, 2485–2487. [CrossRef] [PubMed]

78. Mussali-Galante, P.; Tovar-Sánchez, E.; Valverde, M.; Rojas, E. *Genetic Structure and Diversity of Animal Populations Exposed to Metal Pollution BT—Reviews of Environmental Contamination and Toxicology*; Whitacre, D.M., Ed.; Springer International Publishing: Cham, Germany, 2014; Volume 227, pp. 79–106, ISBN 978-3-319-01327-5.

79. Smith, T.B.; Bernatchez, L. Evolutionary change in human-altered environments. *Mol. Ecol.* **2008**, *17*, 1–8. [CrossRef] [PubMed]

80. Bergland, A.O.; Behrman, E.L.; O'Brien, K.R.; Schmidt, P.S.; Petrov, D.A. Genomic evidence of rapid and stable adaptive oscillations over seasonal time scales in *Drosophila*. *PLoS Genet.* **2014**, *10*. [CrossRef] [PubMed]

81. Acerenza, L. Constraints, trade-offs and the currency of fitness. *J. Mol. Evolut.* **2016**, *82*, 117–127. [CrossRef] [PubMed]

82. Dutilleul, M.; Réale, D.; Goussen, B.; Lecomte, C.; Galas, S.; Bonzom, J.M. Adaptation costs to constant and alternating polluted environments. *Evolut. Appl.* **2017**, *10*, 839–851. [CrossRef] [PubMed]

83. Shirley, M.D.F.; Sibly, R.M. Genetic basis of a between-environment trade-off involving resistance to Cadmium in *Drosophila melanogaster*. *Evolution* **1999**, *53*, 826–836. [CrossRef] [PubMed]

84. Harrison, P.A.; Berry, P.M.; Simpson, G.; Haslett, J.R.; Blicharska, M.; Bucur, M.; Dunford, R.; Egoh, B.; Garcia-Llorente, M.; Geamănă, N.; et al. Linkages between biodiversity attributes and ecosystem services: A systematic review. *Ecosyst. Serv.* **2014**, *9*, 191–203. [CrossRef]

85. Brady, S.P.; Richardson, J.L.; Kunz, B.K. Incorporating evolutionary insights to improve ecotoxicology for freshwater species. *Evolut. Appl.* **2017**, *10*, 829–838. [CrossRef] [PubMed]

86. Szamecz, B.; Boross, G.; Kalapis, D.; Kovács, K.; Fekete, G.; Farkas, Z.; Lázár, V.; Hrtyan, M.; Kemmeren, P.; Groot Koerkamp, M.J.A.; Rutkai, E.; Holstege, F.C.P.; et al. The genomic landscape of compensatory evolution. *PLoS Biol.* **2014**, *12*. [CrossRef] [PubMed]

87. Wan, J.S.H.; Pang, C.K.; Bonser, S.P. Does the cost of adaptation to extremely stressful environments diminish over time? A literature synthesis on how plants adapt to heavy metals and pesticides. *Evolut. Biol.* **2017**, *44*, 411–426. [CrossRef]

88. Palmgren, M.; Engström, K.; Hallström, B.M.; Wahlberg, K.; Søndergaard, D.A.; Sall, T.; Vahter, M.; Broberg, K. AS3MT-mediated tolerance to arsenic evolved by multiple independent horizontal gene transfers from bacteria to eukaryotes. *PLoS ONE* **2017**, *12*, e0175422. [CrossRef] [PubMed]

89. Hendriks, A.J. How to deal with 100,000+ substances, sites, and species: Overarching principles in environmental risk assessment. *Environ. Sci. Technol.* **2013**, *47*, 3546–3547. [CrossRef] [PubMed]

90. Hua, J.; Wuerthner, V.P.; Jones, D.K.; Mattes, B.; Cothran, R.D.; Relyea, R.A.; Hoverman, J.T. Evolved pesticide tolerance influences susceptibility to parasites in amphibians. *Evolut. Appl.* **2017**, *10*, 802–812. [CrossRef] [PubMed]

MDPI

St. Alban-Anlage 66

4052 Basel

Switzerland

Tel. +41 61 683 77 34

Fax +41 61 302 89 18

www.mdpi.com

Water Editorial Office

E-mail: water@mdpi.com

www.mdpi.com/journal/water

Lightning Source UK Ltd.
Milton Keynes UK
UKHW050842070620
364513UK00004B/92